CRACKING IN BUILDINGS

R B Bonshor and L L Bonshor

Construction Research Communications Ltd

Prices for all available BRE publications can be obtained from:
Construction Research Communications Ltd,
151 Rosebery Avenue,
London. EC1R 4QX
Telephone: 0171 505 6622
Facsimile: 0171 505 6606

BR 292
ISBN 1 86081 039 X

© Crown copyright 1996
First published 1996

Published by Construction Research Communications Ltd
by permission of the Controller of HMSO and
the Building Research Establishment.
Applications to copy all or any part of this book
should be made to Construction Research Communications Ltd,
PO Box 202, Watford, Herts, WD2 7QG

Contents

	Page
Preface	v

Part I: THE SCIENCE

Introduction to Part I	2
Chapter 1: The causes of size changes	3
Temperature-induced size changes	3
Moisture-induced size changes	5
Size changes induced by simultaneous temperature and moisture content changes	10
Size changes induced by chemical reactions	10
Sulphate attack	10
Corrosion	11
Carbonation	12
Alkali silica reaction	12
Chapter 2: The mechanism of cracking	13
Chapter 3: Joints as safeguards against cracking	19
Movement joints	19
Assembly joints	21
Inaccuracies in building	21
A remarkable property	23
Tolerances	24
Joints and accuracy	25
Joints and fixings	26

CRACKING IN BUILDINGS Contents

Part II: APPLYING THE SCIENCE

Introduction to Part II — 32

Chapter 4: Temperature-induced size changes — 33
Walls and cladding — 33
    Design principles — 33
    Practical detailing — 35
    Site practices — 39
    Diagnostic principles: walls — 41
    Diagnostic principles: cladding — 41
Flat roofs — 42
    Design principles — 42
    Practical detailing — 43
    Site practices — 44
    Diagnostic principles — 45

Chapter 5: Moisture-induced size changes — 47
Walls — 48
    Design principles — 48
    Practical detailing — 51
    Site practices — 53
    Diagnostic principles — 55
Floors — 56
    Practical detailing for moisture-induced size changes — 56
    Site practices — 57
    Floor finishes — 57

Chapter 6: Chemically-induced size changes — 59
Corrosion of metals — 59
    Wall ties — 59
    Embedded steel in masonry — 60
    Mild steel reinforcing bars in concrete — 61
Sulphate attack — 63
    Fired clay brickwork — 63
    Ground-bearing slabs — 65
Hydration and carbonation — 66
Alkali aggregate reaction — 67
Conversion in high alumina cement concrete — 68

Chapter 7: Cracking due to foundation movements — 69
Design — 69
Diagnosis — 74
    Monitoring movements — 77
Remedial work — 80

Chapter 8: Other causes of cracking — 81
Mechanical damage — 81
Vibration damage — 81
Indirect damage — 82
Frost damage — 83

Appendix 1: Classification of visible damage to walls with particular reference to ease of repair of plaster and masonry — 85

Appendix 2: Crack investigation: a suggested approach — 86

References — 89
Further reading — 92
Index — 93

Preface

Cracks are inevitable in virtually all types of construction because of the kind of materials we use, the ways in which we use or misuse them, and the service conditions which our buildings experience. Nevertheless, cracks are often unsightly and, to the uninitiated, may be a sign of serious problems. Whether a crack is cause for concern or not, of course, depends on circumstances, and whether subsequent action is needed depends on correct diagnosis of the nature of the problem which brought it about.

Accordingly, this book sets out basic information on the science of materials behaviour which is relevant to understanding how and why cracks occur. Given that understanding, much can be done to avoid their occurrence, and to diagnose their cause and repair them so that they do not recur.

It is hoped that readers will find merit in the book in that it collects relevant but scattered information into one source, treats cracking in buildings as a subject in its own right, and provides a systematic approach to whatever is the reader's role in the building business. Its content should therefore be of interest to all who own, occupy, design, build and maintain buildings.

Architects need to design to avoid or, at least, to minimise cracking. They need to be aware of the behaviour of materials and components in response to environmental or other changes, and to be able to assess the consequences of that behaviour for the performance of buildings. The significance of those consequences may determine how much design effort and money should be invested in minimising the risk of cracks developing.

Builders will wish to avoid cracking that might be attributed to their mishandling of materials and components on site (in storage or in the course of construction), to their mistranslation of design requirements or to the quality of their work.

Surveyors undertaking building surveys need to be able to locate and determine the causes of cracks, and to advise on their significance in relation to overall structural integrity and building worth.

Building failure investigators, loss adjusters and expert witnesses in litigation need to consider all possible causes of cracking in buildings so that sound and robust cases can be made for discounting those causes that do not apply and for supporting those that do.

Building owners and maintenance staff wish to be sure that causes of cracking have been correctly identified, and their significance correctly assessed so that time and money are not wasted on unnecessary, irrelevant or, in some cases, even damaging remedial work.

Besides the interest that members in each of the above groups have in relation to their particular role, it is important that they also have a general appreciation of the subject and some understanding of the interests of the other parties in the building process. Surveyors or maintenance staff, for example, will be better equipped to account for a crack in an existing building with knowledge of what designers or builders may or may not have done. The book seeks to meet this generality of interests in two ways.

Firstly, it deals in Part I with the underlying science: the physics (and in some cases the chemistry) underlying the changes of size in materials and components. Part I includes the basic data quantifying size changes and distortions in building materials. Other data include ranges of conditions which are likely to be experienced by parts of buildings in service and which determine the size changes occurring in particular circumstances. These data are essential both in designing to avoid cracking damage and in the diagnosis of the causes of damage in existing structures. The Part describes the mechanisms by which the size changes potentially produce intolerable strain, and consequent distortion or cracks. It deals also with the way in which unavoidable inaccuracies in building construction modify or negate the design provisions made to accommodate changes of size in components and structures. Thus Part I provides essential and fundamental information relevant to all, whatever their role in building.

Secondly, in Part II, it deals with the causes of cracking covered in principle in Part I but setting them in real building contexts, taking each building element in turn. Here the common interests of the various parties are met by presenting the information in a common format, typically:

- design principles
- practical detailing
- site practices
- diagnostic principles
- remedial work or repairs

Under 'design principles', the factors operating in each case are identified. Under 'practical detailing', design solutions are described for particular cases. The 'site practices' section deals with the ways in which site activities influence subsequent behaviour of materials and components in service. The 'diagnostic principles' section explains what factors must be present for any particular conclusion about causes to be valid – and how to confirm their presence. (Appendix 2 shows a suggested approach to crack investigation.) Finally, under 'remedial work or repairs', the need for action and its nature are described.

Thus the interests of all roles are brought together in each successive package of information.

All of the information presented in both Part I and Part II already exists elsewhere, though scattered among a considerable number of different sources. But the information is not only scattered; much of it appears under headings that do not give an immediate impression that the content might be relevant to cracking or distortion – alkali aggregate reaction and re-covering old timber roofs, for example – so that it might easily be missed in a library search for information and guidance on the causes and consequences of cracking. The extensive bibliography provided should also help in this respect.

In the United Kingdom, there are three separate sets of building regulations: for England and Wales, Scotland, and Northern Ireland. There are many common provisions between the three sets, but there are also differences. The fact that references to building regulations are to those for England and Wales should not make the book inapplicable to Scotland and Northern Ireland.

One aspect of cracking in buildings is intentionally omitted. The design of structural members to control cracking under service loads, or under handling stresses, is both too specialised and too well covered in books on structural design to warrant it being included here. Nevertheless, there should be sufficient information in this book for readers to distinguish between cracking due to service loads and to other causes.

We are immensely grateful to the following members and former members of BRE staff who have contributed to, or commented on, the preparation of this book: R N Cox, Dr Nora Crammond, Dr R C de Vekey, R M C Driscoll, M A Halliwell, H W Harrison, Dr P J Nixon. Also to Dr A J Wadge of the British Geological Survey for his advice on the content of the tables on sources of information and methods of investigation relating to topography, vegetation, drainage and ground conditions.

R B and L L Bonshor

Part I

THE SCIENCE

Introduction to Part I

In most artefacts a crack indicates that the item has failed – or will do so shortly, no matter whether that item is a turbine blade or a teacup handle – and that urgent repair or replacement is essential.

Cracking in buildings does not follow this general conception. The total collapse of a building may indeed be preceded by an observable, apparently innocuous, hairline crack in its fabric; but total or even partial collapse of a building within its expected service life is fortunately rare indeed, barring acts of war, earthquake and similar catastrophic events.

Virtually all parts of buildings are subjected to continuing size changes, expanding or perhaps contracting as the materials used in their construction respond to changes in temperature or moisture content. Buildings are comparatively large, complex and rigid structures, constructed from disparate materials with component parts subjected simultaneously to differing environmental conditions. It is not surprising that cracks are inevitable, though only some impair the serviceability of a building or may do so if they widen further. (Appendix 1 presents a method of classifying visible damage to walls.) Such cracks may justify repair or require measures to ensure that they do not develop further. Distinguishing these from the remainder, the vast majority, requires an adequate understanding of the various factors involved: the materials technology, the causes, the mechanisms and the performance consequences of cracks. To that extent one of the aims of this book is to discourage any automatic assumption that a crack is necessarily significant – diminishing the building's integrity and worth, and demanding urgent remedy – and to substitute over-reaction with calm and reasoned appraisal based on sound knowledge.

Chapter 1
The causes of size changes

Building materials respond to changes in the conditions to which they are subjected. These responses are commonly referred to as movements – thermal movements and moisture movements, for example. They are better described as size changes in order to distinguish them from true movements such as occur when components are physically displaced in position; for example by changes of load or support. Both movements and size changes can cause cracks. Movements (in the sense of physical displacements) are dealt with in Chapter 7 in the context of support changes.

A further reason for making a clear distinction between size changes and movements is that the term moisture movement is commonly used both for change of size resulting from change in moisture content and to describe the movement of moisture through materials.

It is particularly important to recognise that size changes are occurring virtually continuously (in response, for example, to continuously changing temperatures) regardless of whether or not other disturbing forces may produce movements at any time in the life of a building.

This chapter brings together the available information about:

- changes in conditions experienced by building materials and components
- the magnitude of such changes in conditions
- the magnitude of the corresponding size changes

All such size changes can result in cracking or contribute to, or modify, other causes of cracking. Buildings are constructionally complex with different parts made from differing materials and subjected to differing conditions. Often different causes of size change are operating simultaneously, even in the same component. Such complexity is resolved by considering each kind of condition change in turn.

Temperature-induced size changes

All building components, whatever the material from which they are made, change in size as a consequence of changes in their temperature. (The terms temperature-induced size change and thermally-induced size change are used widely in building science and have virtually the same meaning, but, for the purpose of consistency, the former expression will be used throughout this book.) The subject is best dealt with on the basis of an initial assumption that the material of which a component is made is homogeneous, that all parts of it experience the same change of temperature, and that no restraints are imposed, whether from within the component or from outside it.

In many cases this assumption, though not strictly correct, will be adequate in practice. In some cases the assumption will not alone be adequate: temperature distribution or the homogeneity of the material may be far from uniform and size change will be accompanied by significant distortion. But, in all cases, consideration of the unrestrained linear change of size is still the first step in an analysis of distortion.

On the basis of that initial assumption, calculation of the size change accompanying a temperature change requires only three pieces of information:

- the coefficient of linear thermal expansion α of the material concerned
- the size L of the dimension concerned
- the extent t of the temperature change

The size change R is then given by:

$$R = \alpha \, L \, t$$

The equation is obviously a simple one to use, but a substantial error will be made unless care is taken to keep the decimal point in the right place, remembering that thermal coefficients are usually in the form of a value multiplied by 10^{-6}.

In some cases – for example, a designer considering various possible choices of materials to specify, or a surveyor considering various possible temperatures that may have been experienced by the structure – the nomogram in Figure 1.1 (on page 4) will be found more convenient than repetitive calculation.

CRACKING IN BUILDINGS The causes of size changes

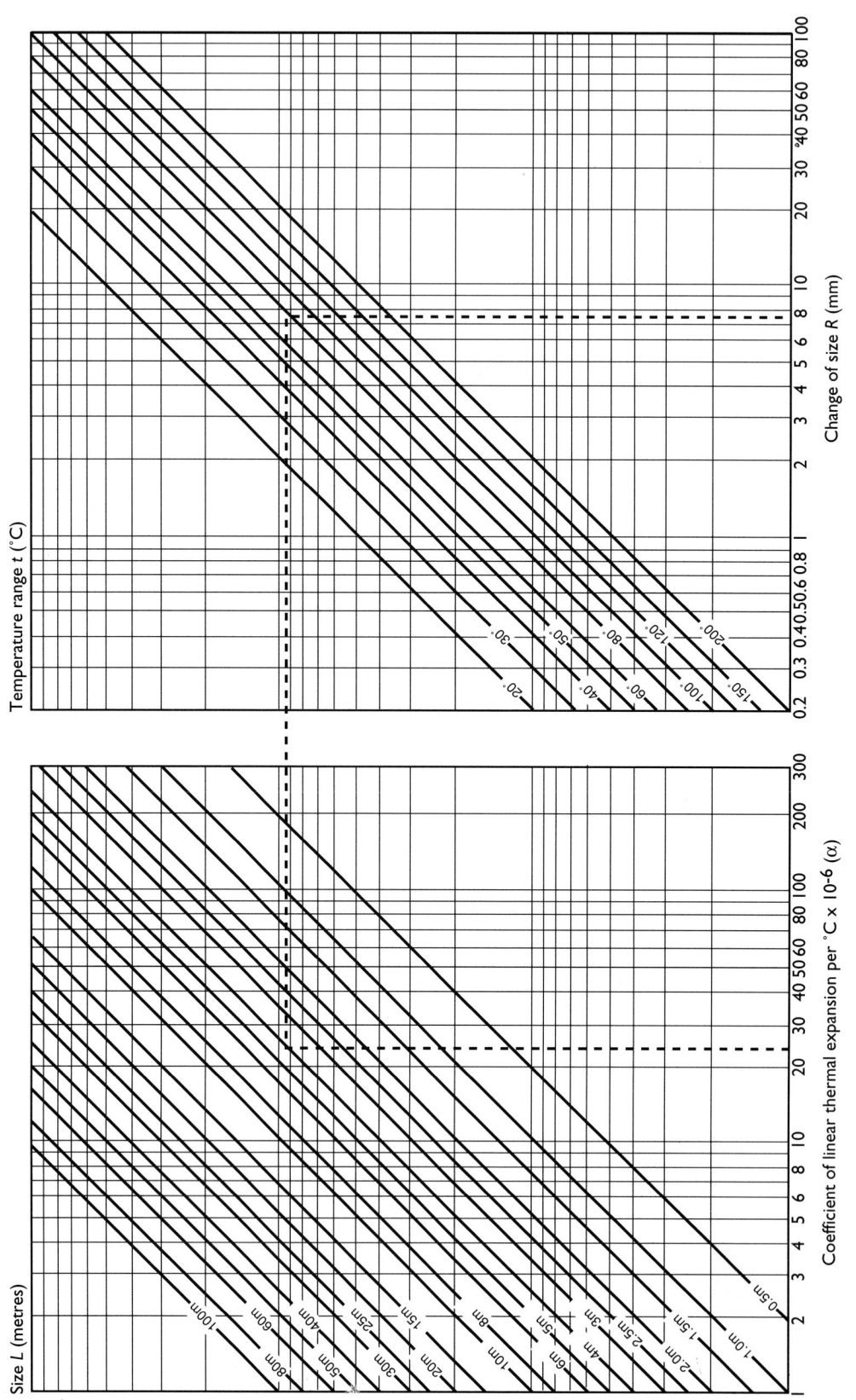

Figure 1.1
Computation of thermal expansion

Table 1.1 (on page 6) provides coefficients of linear thermal expansion for a wide variety of common building materials. An example of their use in a calculation is given in the feature panel at the foot of this page.

Chapter 2 describes how changes in size may be modified by restraints – imposed, for example, by rigid fixing at both ends of the component – and Chapter 4 explains how the results of calculations of this kind can be applied to practical circumstances.

Thus far it has been assumed that the temperature range experienced by building materials is a known quantity in any particular case. In a case in which there is dispute about the cause of a failure, it may be necessary to measure directly the temperatures of materials in the building. But most often what is needed is an indication of the order of magnitude of the temperature range so that reasonably representative values for t can be selected. Unfortunately, no comprehensive authoritative list exists (such as might have been obtained by long term nationwide measurements) of temperatures of the fabric of buildings of all kinds in the UK.

Table 1.2 (on page 7) provides such information as there is, in general terms, about temperature ranges of materials in service in the UK, and users will have to interpret this in terms of the orientation, the exposure, insulation or shading of the parts concerned and in the light of information about the daily or seasonal range of ambient air temperatures. The daily or seasonal temperature range will provide some indication of the smallest range likely to be experienced by exterior parts.

Thus the information in Tables 1.1 and 1.2 enables estimates to be made of the unrestrained linear response of single-material components to temperature changes. But, temperature is not the only factor to take into account. Very many building materials are also affected dimensionally by moisture content changes.

Moisture-induced size changes

All temperature-induced size changes are reversible, though more often described as cyclic. However, size changes in response to changing moisture conditions may be either reversible or irreversible in character. With the exception of fired clay products (and discounting chemical phenomena which are described in Chapter 6), all irreversible size changes are negative; in other words, the components experience irreversible shrinkage.

However, the irreversible shrinkage – for example, which characterises cement-based products of all kinds – does not of course preclude the component from experiencing reversible size changes induced by increasing and decreasing moisture content. Similarly, fired clay products are also subject to reversible size changes with changing moisture content.

Once again the subject is best dealt with in terms of an initial assumption of unrestrained and uniform linear response. Moisture-induced size changes, both reversible and irreversible, are usually given as simple percentages. Thus calculation of a size change accompanying a moisture content change requires only two items of information:

- the percentage value
- the size L of the dimension concerned

The size change R is then given by:

$R = L \times \%$ value

Table 1.3 (on page 8) provides values for both reversible and irreversible moisture-induced changes. An example of their use in a calculation is given in the feature panel on page 9. The modifying effects of restraint are considered on page 13.

AN EXAMPLE OF TEMPERATURE-INDUCED SIZE CHANGE

Suppose that a 3 m (3000 mm) length of, say, an aluminium alloy mullion in curtain walling is subjected to a seasonal change of temperature over a range from −20 °C to 50 °C, as it might well be in practice. Thus the temperature range t is 70 °C. The thermal coefficient α for aluminium alloy, found from Table 1.1, is 24×10^{-6} per °C. The size change R in the 3000 mm length L is given by:

$R = \alpha L t$
$= 24 \times 10^{-6} \times 3000 \times 70$
$= 0.000024 \times 3000 \times 70$
$= 0.24 \times 3 \times 7$
$= 5.04$ mm

So if the mullion is fixed rigidly at its lower end only, the joint between its top and the bottom of the next mullion above (similarly fixed at its lower end) will change in size by 5 mm.

CRACKING IN BUILDINGS The causes of size changes

TABLE 1.1 THERMAL EXPANSION COEFFICIENTS

Note: unless more specific data are available, design should be based on the higher value where a range is shown

Material	Coefficient of linear thermal expansion α per °C $\times 10^{-6}$	Material	Coefficient of linear thermal expansion α per °C $\times 10^{-6}$
Natural stones		**Wood and wood laminates***	
Granite	8–10	Softwoods	4–6 with grain
Limestone	3–4		30–70 across grain
Marble	4–6	Hardwoods	4–6 with grain
Sandstone	7–12		30–70 across grain
Slate	9–11	Plywood	†
		Blockboard and laminboard	†
Cement based composites			
Mortar and fine concrete	10–13	**Woodchip and fibrous materials**	
Dense aggregate concrete:		Hardboard	†
gravel aggregate	12–14	Medium board	†
crushed rock (except limestone)	10–13	Softboard	†
limestone	7–8	Chipboard	†
Steel fibre reinforced concrete	5–14	Wood-wool cement	†
Aerated concrete	8		
Lightweight aggregate concrete:		**Rubbers and plastics, etc**	
medium lightweight	8–12	Asphalt	30–80
ultra lightweight (exfoliated vermiculite and		Pitch fibre	40
expanded perlite)	6–8	Ebonite	65–80
Asbestos cement	8–12	Thermoplastics:	
Glass reinforced cement	7–12	PVC, PVC-U and PVC-C	40–70
		polyethylene (low density)	160–200
Calcium silicate based composites		polyethylene (high density)	110–140
Asbestos wallboard and substitutes	5–12	Polypropylene	80–110
Asbestos insulating board and substitutes	2.5–7.2	Polycarbonate	60–70
		Polystyrene	60–80
Gypsum and gypsum based composites		Acrylic	50–90
Dense plasters; plasterboard	18–21	Acetal	80
Sanded plasters	12–15	Polyamide	80–130
Lightweight plasters	16–18	ABS	60–100
Glass reinforced gypsum	17–20	Thermosets (laminates):	
		phenol and melamine formaldehyde	30–45
Brickwork, blockwork and tiling		urea formaldehyde	27
Concrete brickwork and blockwork:		Cellular (expanded):	
dense aggregate	6–12	PVC	35–50
lightweight aggregate (autoclaved)	8–12	phenolic	20–40
aerated (autoclaved)	8	urea formaldehyde	30–90
Calcium silicate brickwork	8–14	polyurethane	20–70
Clay or shale brickwork or blockwork	5–8	polystyrene	15–45
Clay tiling	4–6	Reinforced:	
		GRP (chopped strand)	20–35
Metals		carbon fibre (orientated):	
Cast iron	10	parallel to reinforcement	0–0.05
Plain carbon steel	12	perpendicular to reinforcement	30–70
Stainless steel:			
austenitic	18	**Glass**	
ferritic	10	Plain, tinted and opaque	9–11
Aluminium and alloys	24	Foamed (cellular)	8.5
Copper	17		
Bronze	20		
Aluminium bronze	18		
Brass	21		
Zinc:	33 parallel to rolling		
	23 perpendicular to rolling		
Lead	30		

* More specific data can be found in: *The strength properties of timbers* by Gwendoline M Lavers, BR 241. Garston, Construction Research Communications Ltd, 1983; ISBN 0 85125 562 0. *A handbook of softwoods*, SO 39. London, HMSO, 1977; ISBN 0 11470 563 1. And *Handbook of hardwoods*, SO 7. London, HMSO, 1972; ISBN 0 11470 541 0.

† No data available

The causes of size changes CRACKING IN BUILDINGS

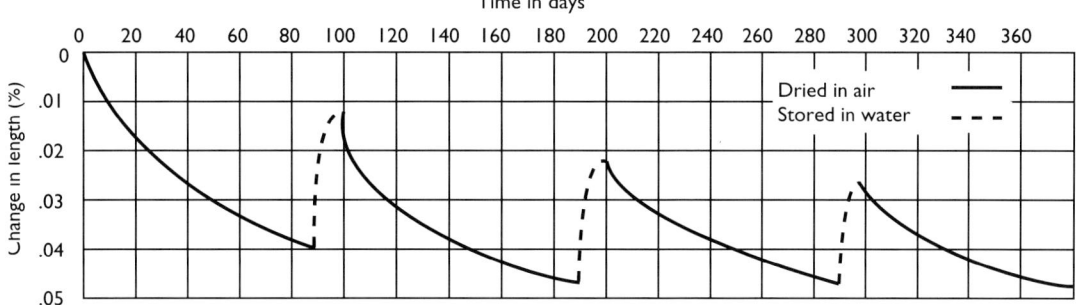

Figure 1.2
Change in length of concrete specimens (over successive 90 day periods of wetting and drying) from initial cast length

TABLE 1.2 EXAMPLES OF SERVICE TEMPERATURE RANGES OF MATERIALS (VALID FOR UK ONLY)

Notes: The following situations are not included in the examples below and may give rise to temperature extremes more severe than those listed:
- dark surfaces under glass (eg, solar collectors)
- materials used in cold rooms or refrigerated stores
- materials used for, or in proximity to, heating, cooking and washing appliances, or flues and heat distribution networks

	Min (°C)	Max (°C)	Range (°C)
External			
Cladding, walling, roofing:			
heavyweight:			
light colour	−20	50	70
dark colour	−20	65	85
lightweight, over insulation:			
light colour	−25	60	85
dark colour	−25	80	105
Glass:			
coloured or solar control	−25	90	115
clear	−25	40	65
Freestanding structures or fully exposed structural members:			
concrete:			
light colour	−20	45	65
dark colour	−20	60	80
metal:			
light colour	−25	50	75
dark colour	−25	65	90
Internal			
Normal use	10	30	20
Empty/out of use	−5	35	40

CRACKING IN BUILDINGS The causes of size changes

TABLE 1.3 MOISTURE-INDUCED SIZE CHANGES

Note: unless more specific data are available, design should be based on the higher value where a range is shown

Material	Reversible moisture movement (%)	Irreversible moisture movement (+) expansion (−) contraction (%)
Natural stones		
Granite		
Limestone	0.01	
Marble		
Sandstone	0.07	
Slate		
Cement based composites		
Mortar and fine concrete	0.02–0.06	0.04–0.10 (−)
Dense aggregate concrete:		
gravel aggregate	0.02–0.06	0.03–0.08 (−)
crushed rock (except limestone)	0.03–0.10	0.03–0.08 (−)
limestone	0.02–0.03	0.03–0.04 (−)
Steel fibre reinforced concrete	0.02–0.06	0.03–0.06 (−)
Aerated concrete	0.02–0.03	0.07–0.09 (−)
Lightweight aggregate concrete:		
medium lightweight	0.03–0.06	0.03–0.09 (−)
ultra lightweight (exfoliated vermiculite and expanded perlite)	0.10–0.20	0.20–0.40 (−)
	(Values for plain concrete; moisture movements may be partially restrained by appropriately placed reinforcement)	
Asbestos cement	0.10–0.25	0.08 (−)
Glass reinforced cement	0.15–0.25	0.07 (−)
Calcium silicate based composites		
Asbestos wallboard and substitutes	0.14–0.27	
Asbestos insulating board and substitutes	0.16–0.25	
Brickwork, blockwork and tiling		
Concrete brickwork and blockwork:		
dense aggregate	0.02–0.04	0.02–0.06 (−)
lightweight aggregate (autoclaved)	0.03–0.06	0.02–0.06 (−)
aerated (autoclaved)	0.02–0.03	0.05–0.09 (−)
Calcium silicate brickwork	0.01–0.05	0.01–0.04 (−)
Clay or shale brickwork or blockwork	0.02	0.02–0.10 (+)
Clay tiling	†	†
Wood and wood laminates*		
Softwoods	Negligible with grain. Across grain: 0.6–2.6 tangential (1), 0.45–2.0 radial (1)	
Wood and wood laminates* (cont)		
Hardwoods	Negligible with grain. Across grain: 0.8–4.0 tangential (1), 0.5–2.5 radial (1)	
Plywood	0.15–0.20 with grain (2), 0.20–0.30 across grain (2)	
Blockboard, laminboard	0.05–0.07 with core (2), 0.15–0.35 across core (2)	
Woodchip and fibrous materials	(On length or width, values for thickness may be up to 30 times greater)	
Hardboard	0.30–0.35 (2)	
Medium board	0.30–0.40 (2)	
Softboard	0.40 (2)	
Chipboard	0.35 (3)	
Wood-wool cement	0.15–0.30 on length, 0.25–0.40 on width	
Rubbers and plastics, etc		
Asphalt		
Pitch fibre	0.2–0.3	
Ebonite		
Thermoplastics:		While not subjected to moisture effects, some plastics may be liable to irreversible progressive contraction due to loss of volatiles and related causes
PVC, PVC-U and PVC-C		
polyethylene (low density)		
polyethylene (high density)		
Polypropylene		
Polycarbonate		
Polystyrene		
Acrylic		
Acetal		
Polyamide		
ABS		
Thermosets (laminates):		
phenol and melamine formaldehyde		
urea formaldehyde		
Cellular (expanded):		
PVC		
phenolic		
urea formaldehyde		
polyurethane		
polystyrene		
Reinforced:		
GRP (chopped strand)		
carbon fibre (orientated)		

(1) Based on 60% and 90% relative humidities
(2) Based on 33% and 90% relative humidities
(3) Based on 65% and 90% relative humidities
* More specific data can be found in: *The strength properties of timbers* by Gwendoline M Lavers, BR 241. Garston, Construction Research Communications Ltd, 1983; ISBN 0 85125 562 0. *A handbook of softwoods*, SO 39. London, HMSO, 1977; ISBN 0 11470 563 1. And *Handbook of hardwoods*, SO 7. London, HMSO, 1972; ISBN 0 11470 541 0.
† No data available

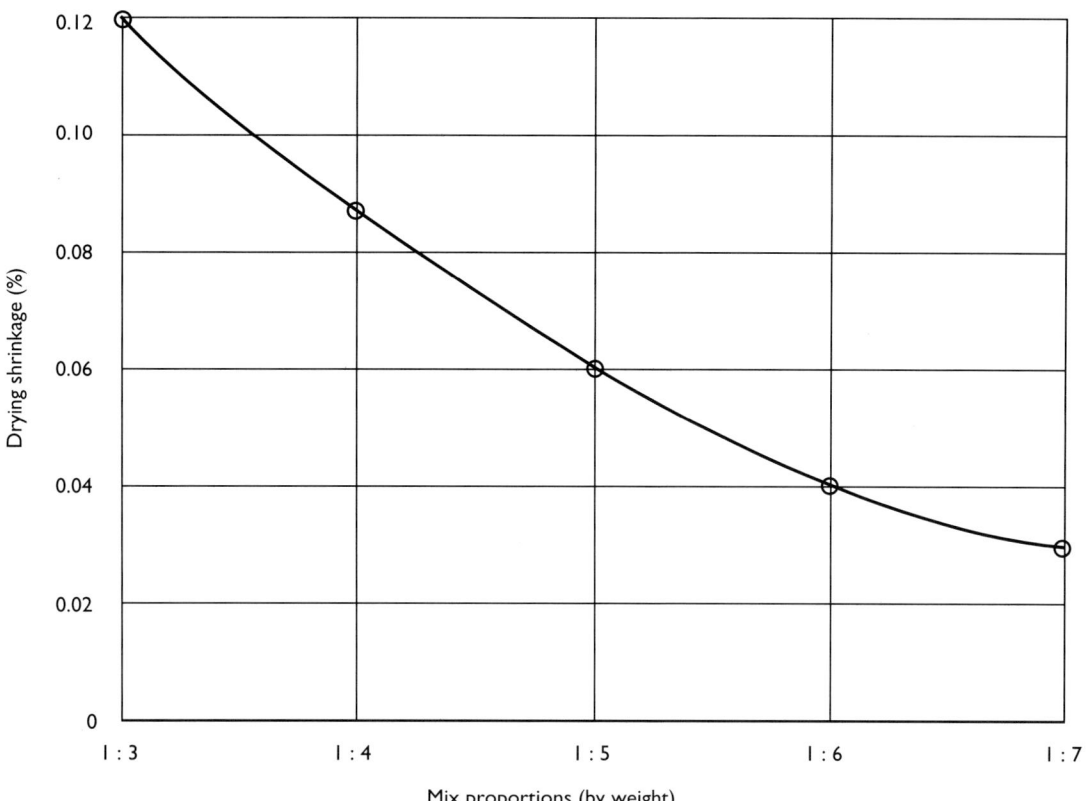

Figure 1.3
Variation of drying shrinkage with cement content of concrete or mortar: drying shrinkage of specimens (water:cement ratio 0.5) after 6 months in air at 20 °C and 50% relative humidity

---

AN EXAMPLE OF MOISTURE-INDUCED SIZE CHANGE

Suppose that a 3 m (3000 mm) length of a component made from unreinforced gravel aggregate concrete experiences the full range of moisture content change likely to occur in service – from Table 1.3, 0.06%. The corresponding change of size is given by:

$$R = L \times \% \text{ value}$$
$$= \frac{3000 \times 0.06}{100}$$
$$= 1.8 \text{ mm}$$

It should be noted that the calculated size change is that which would occur over the full range of change in moisture content. If the moisture content cycles about some value within the range, the total of both positive and negative size changes would not exceed 1.8 mm.

If L is the length as initially cast, account can be taken of the initial irreversible size change (shrinkage in this case). From Table 1.3 the relevant value is –0.08%:

$$R = L \times \% \text{ value}$$
$$= \frac{3000 \times -0.08}{100}$$
$$= -2.4 \text{ mm}$$

Thus the component in this example would be expected to shrink from its initial as-cast length by 2.4 mm and thereafter its size might change reversibly, in response to moisture content changes, over a range of 1.8 mm (say 2 mm).

The values given in Table 1.3 for reversible moisture-induced size changes are based on the extreme moisture contents likely to occur in normal use. The range for the inorganic products listed is that for extreme wet (but not saturated) to extreme dry external exposure. For the wood and wood-based products listed, the ranges correspond to equilibrium moisture content at the two extremes of relative humidity given at the foot of the Table. Reversible moisture-induced size changes can be reduced, of course, from the values in Table 1.3 by limiting the range of change of moisture content. Chapter 5 deals with the practical steps that can be taken.

As Figure 1.2 (on page 7) shows, reversible moisture-induced size changes can be superimposed on initial drying shrinkage. The amount of initial shrinkage in concrete or mortar mixes varies with cement content. Figure 1.3 (on page 9) shows that a 1:4 mix produces more than twice the shrinkage of a 1:6 mix. A general rule, therefore, is that consequential cracking will be much reduced if the leanest mixes consistent with adequate strength are used.

Size changes induced by simultaneous temperature and moisture content changes

Many building materials are subjected to simultaneous temperature and moisture content changes. When these changes arise from separate and independent causes, a reasonable approximation is to treat them as additive according to their sign. However, when the causes are interdependent – moisture content falling, for instance, as temperature rises – the combined effect is strongly influenced by the rate of change of conditions. For example, response to temperature changes occurring within a daily range can generally be assumed to be more rapid than associated moisture-induced size changes – that is to say, thermal response tends to dominate in the short term. But no generally applicable guidance exists and design must be a matter of judgement, erring on the safe side in the assumptions made.

Size changes induced by chemical reactions

Size changes accompanying chemical reactions cannot be quantified and tabulated: they differ in this respect from the size changes due to temperature and moisture, and they continue for as long as conditions remain favourable to the reaction and further reactive materials remain available. These changes are therefore dealt with here in a different way, providing descriptive rather than numerical information. In any case, the objective for the building designer is to avoid their occurrence rather than to accommodate them. For the surveyor too, the interest is not in the amount of size change but in the characteristic consequences of different reactions so that they can be distinguished from each other and a diagnosis made. The information given here is therefore intended to be no more than is sufficient to enable designers and builders to have sufficient awareness of the underlying science to be able to avoid problems, and for surveyors and failure investigators to recognise them.

Chemical reactions that potentially produce cracks include sulphate attack, corrosion, carbonation, and alkali silica reaction.

Sulphate attack

All materials based on ordinary Portland cement (OPC) are potentially vulnerable but, for attack to occur, three constituents are necessary and the removal or absence of any one of them ensures freedom from attack:

- soluble sulphate salts
- OPC
- persistent wetness

The reaction is accompanied by expansion. In brickwork the effect is most pronounced in the bed joints: the mortar becomes crumbly, with horizontal cracks; at the boundaries of affected areas there may be white crystalline deposits. Rendered brickwork will show horizontal cracks matching bed joint locations. The extent of cracking will relate to the distribution of water in the wall – more, and wider, cracks in the wetter parts, usually the topmost courses. Soldier courses and brick-on-edge copings are particularly vulnerable (unless the bricks have been selected for their low sulphate content) due to the relatively larger number of perpends in which attack and expansion can occur. Brick-on-edge sills can also be vulnerable for the same reason. Brick parapet walls, particularly if rendered, are vulnerable because they are likely to be persistently wet.

The forces developed are large. An affected boundary wall may punch in the corner of an adjacent building or disrupt an arched gateway. Alternatively the expansive stress may be relieved by upward bowing of the upper courses of a freestanding wall or by outward bowing (either in the height or on plan) of a wall that is contained by a rigid structure. The expansive forces may cause walls to oversail the damp proof course (DPC), distort salient corners or disrupt associated detailing. Freestanding walls and the brick outer leaves of cavity walls that are persistently wetter on one face than on the other may bow – an effect commonly apparent also in unlined chimney stacks in which water vapour in the flue gases condenses persistently and preferentially on the colder, wetter side, inducing greater expansive attack on that side.

Sulphate attack can occur also in concrete ground-supported slabs laid directly on fill containing sulphates; the resulting expansion at the lower face of the slab produces upward doming, tension cracks in the upper surface and outward displacement of containing walls. Concrete piles and footings can similarly be attacked by sulphate-bearing groundwater.

The design objective of avoiding attack is achieved by using materials low in soluble sulphates or by isolating cement-based products from sulphate-bearing materials, by keeping potentially affected parts dry or by using sulphate-resisting cement in mortar and concrete.

For surveyors and failure investigators it is important to be able to recognise the characteristic consequences as described and to check that the three essential constituents are indeed present or clearly have been at some time. Full confirmation of the diagnosis can be obtained by specialist examination of affected cement-based material for the presence of ettringite or thaumasite, the end-products of the reaction.

Corrosion
Corrosion of metals is relevant in the present context only insofar as it is accompanied by expansion, and so may produce cracks in structures. In this limited context the material of most concern is mild steel. The corrosion product of mild steel potentially occupies several times the volume of the parent metal and the expansive forces developed are considerable. Normal atmospheric corrosion of iron and mild steel requires only two constituents:

- air
- water

but the exclusion of either constituent avoids the problem.

In practice, however, the corrosion of mild steel in buildings is influenced by several other factors which fall into two groups:

- those that protect against corrosion (eg, paint films, sacrificial protection, and an alkaline environment)
- those that accelerate corrosion (eg, acid atmospheric pollutants, and chloride ions from additives or from a marine environment)

The protection of steel components by coatings of many different kinds is dealt with very extensively in BS 5493[1] and, for example, in such specific standards as BS 4147[2]. These standards enable environmental exposure to be taken into account and protective systems to be selected according to life requirements. Designers, both of buildings and of steel components manufactured for use in buildings, should be aware of the comprehensive recommendations made.

In the context of cracking, it is the corrosion of steel which is incorporated in some way into materials or structures that is of first concern since associated expansion is then most likely to disrupt other parts. Wall ties are one example. The earliest examples of wall tie corrosion occurred in exposed masonry built in black ash (and therefore acidic) mortar. The ultimate amount of corrosion product depends upon the amount of parent metal available to corrode and therefore the earliest observed failures involved strip ties rather than wire ties. Corrosion of that part of the tie that is embedded in the outer leaf forces open the bed joints in which the tie is placed. Thus the distinguishing characteristic is that cracks occur in bed joints at vertical tie-spacing intervals (compare this with the case of sulphate attack, in which all bed joints in the wetter areas are affected). Wire ties appear to have insufficient corrodible volume to produce visible cracks in bed joints when corrosion occurs, though their effectiveness as ties may be impaired or destroyed. Increased coating-weight requirements in standards and stopping the use of black ash mortar should greatly reduce the incidence of cracking from this cause, though surveyors will continue to encounter problems with buildings that pre-date these changes.

The corrosion of mild steel reinforcement is affected by rather different factors. The alkali liberated by the hydration of cement provides a favourable environment for the steel, inhibiting corrosion. But atmospheric carbon dioxide, diffusing into the concrete, reacts with the alkali, converting it into carbonate and so reducing progressively the alkaline protection afforded to the steel. The rate at which this occurs is dependent primarily on the porosity, and the thickness, of the surrounding concrete. Cracking and corrosion of reinforcement is dealt with more fully in BRE Digest 389[3].

For the designer, therefore, the aim is to provide adequate thickness of concrete cover and to

CRACKING IN BUILDINGS The causes of size changes

specify concrete mixes that can be fully compacted with the lowest water:cement ratio compatible with full compaction. Reinforced concrete members should also be designed to minimise cracking under load, since cracks provide routes by which carbon dioxide can gain deep access to the interior of the member. Carbonation will still proceed, albeit slowly, in the densest concrete unless some means (such as an impermeable paint film) can be employed to exclude carbon dioxide totally.

When reinforcing steel corrodes, the accompanying expansion cracks the concrete around it. For surveyors, the earliest signs are likely to be rust stains in which a pattern matching the location of stirrups (ie, the steel with least cover) may be apparent; if corrosion is further advanced the pattern of the main bar locations may be visible in the distribution of cracks and there is likely to be some spalling and detachment of surface concrete in these areas. It should be noted, though, that isolated rust-coloured stains without associated cracks may not be an early indication of corrosion but of the presence of pyrites in the concrete aggregate used. Rust staining will not always occur with carbonated concrete; it is more likely to occur where chlorides are present in concrete.

Carbonation

The reaction of atmospheric carbon dioxide with alkali in concrete has already been mentioned in the context of steel corrosion. The reaction can arise in other cement-based materials. It is accompanied by shrinkage and, in thin sheet materials such as fibre-reinforced cements, can produce marked distortion if accessibility to atmospheric carbon dioxide is not uniform: as when a sheet is painted on one side only or where coatings on opposite faces have different permeability to carbon dioxide. In such cases, if the sheets are firmly fixed, the restraint offered to distortion can readily crack the sheets. Calcium silicate-based sheets and boards may also contain appreciable free lime, and the safest course is to ensure that any coatings provide equal permeability on both faces by coating both faces or neither.

For the surveyor or failure investigator, the origin of the problem will be apparent from the fact that the material, as installed, is more accessible to carbon dioxide on one face than on the other and from the fact that distortion in the form of bowing or doming has occurred: the face which is better protected from carbon dioxide being the convex face. It may be necessary to unfasten fixings that are restraining distortion before the distortion can be seen. Full confirmation will require laboratory investigation to establish the presence of free, as yet uncarbonated, lime.

Alkali silica reaction

Alkali silica reaction (ASR) is the most common form of alkali aggregate reaction. It is an expansive reaction that, in unrestrained concrete, produces map-pattern cracking very similar to the cracking produced by initial drying shrinkage. Fortunately it is comparatively rare. For the reaction to occur, four constituents are necessary:

- high cement content
- high alkali content in the cement
- exposure to water
- reactive aggregate

Where ASR may occur, special design and specification measures are necessary. These measures go beyond the scope of this book and designers are referred to The Concrete Society's Technical Report No 30[4].

For surveyors and failure investigators the main problem is to distinguish ASR cracking from initial drying shrinkage cracking. Generally the latter occurs within the first year of the structure's life whereas ASR cracking has not been diagnosed in buildings less than five years old. Dating the first occurrence of cracking in a particular building may of course still be difficult for the surveyor. Frost damage can also produce similar cracking, though it can usually be distinguished from ASR by the presence also of surface spalling. A further difficulty is that where reinforced concrete members suffer ASR, cracks tend to run parallel to main bars in much the same way as do cracks caused by corrosion of the steel; indeed ASR cracks may promote corrosion of the steel (by admitting atmospheric carbon dioxide) as well as leaving the concrete more vulnerable to frost damage. The only certain confirmation of a diagnosis of ASR is by way of microscopical examination; for example, of core samples in the laboratory.

Chapter 2

The mechanism of cracking

Between the fundamental cause of a crack and the crack itself there is always a mechanism – a physical process by which changes of size in a material or component are translated into displacement, or into stress greater than the material or component can withstand. Such displacement or stress may also be transmitted to neighbouring materials or components. Figure 2.1 (on page 14) illustrates some simple examples of what is meant by mechanism.

In practice the paths taken by the forces developed by size changes depend on the way components are located relative to each other; they also depend on the location and effectiveness of restraints such as those provided by mechanical fixings, adhesion and friction.

A building designer considering what may be the consequence of a particular change of size in service is, in effect, trying to trace the path likely to be taken by the force developed – and trying to predict what that force will do, physically, to the construction. In short, the designer looks for the particular mechanism by which the size change may produce damage.

Usually an inspection of a proposed design arrangement and its detailing is enough. The path of the developed force will be apparent from the juxtaposition of the components and its ultimate effect predictable from the calculated amount of size change and from past experience of the behaviour of similar construction. Often the design will be such that change of size in a component is fully restrained (or is best assumed to be so), whether unavoidably or intentionally. The component will then be subject to stress which will need to be evaluated in order to determine whether it is tolerable or whether damage – commonly cracking – will result.

Stress is the product of the modulus of elasticity E for the material and strain:

Stress = $E \times$ strain

Strain is the change in length divided by the length. Values of E for common building materials are given in Table 2.1 (on page 15).

Since thermal coefficients are expressed also in terms of change in length per unit length, it follows that for temperature-induced size changes:

Strain = $\alpha\, t$

where α is the coefficient of linear thermal expansion and t is the temperature range experienced. In this case, therefore:

Stress = $E \times \alpha\, t$

Similarly, for moisture-induced size changes:

Stress = $\dfrac{E \times \%\ \text{value}}{100}$

If stresses so calculated exceed, for example, the ultimate tensile strength of the material, then tensile cracking will occur. Whatever form of cracking is promoted by the restraint, there will be a corresponding force acting on the restraining construction:

Force = stress × area
= $E \times \alpha\, t \times A$
(for temperature-induced size changes)
or = $\dfrac{E \times \%\ \text{value}}{100} \times A$
(for moisture-induced size changes)

Thus the effect of that force, in terms of the stress it also induces in the restraining construction, should be considered in order to check whether that construction also may suffer damage.

If the restrained component is relatively thin and flexible, total restraint of what would otherwise have been a linear expansive size change is likely to produce bow in the component. A designer may well wish to set a limit on the amount of bow that is acceptable and to calculate the amount of bow accruing from the

CRACKING IN BUILDINGS The mechanism of cracking

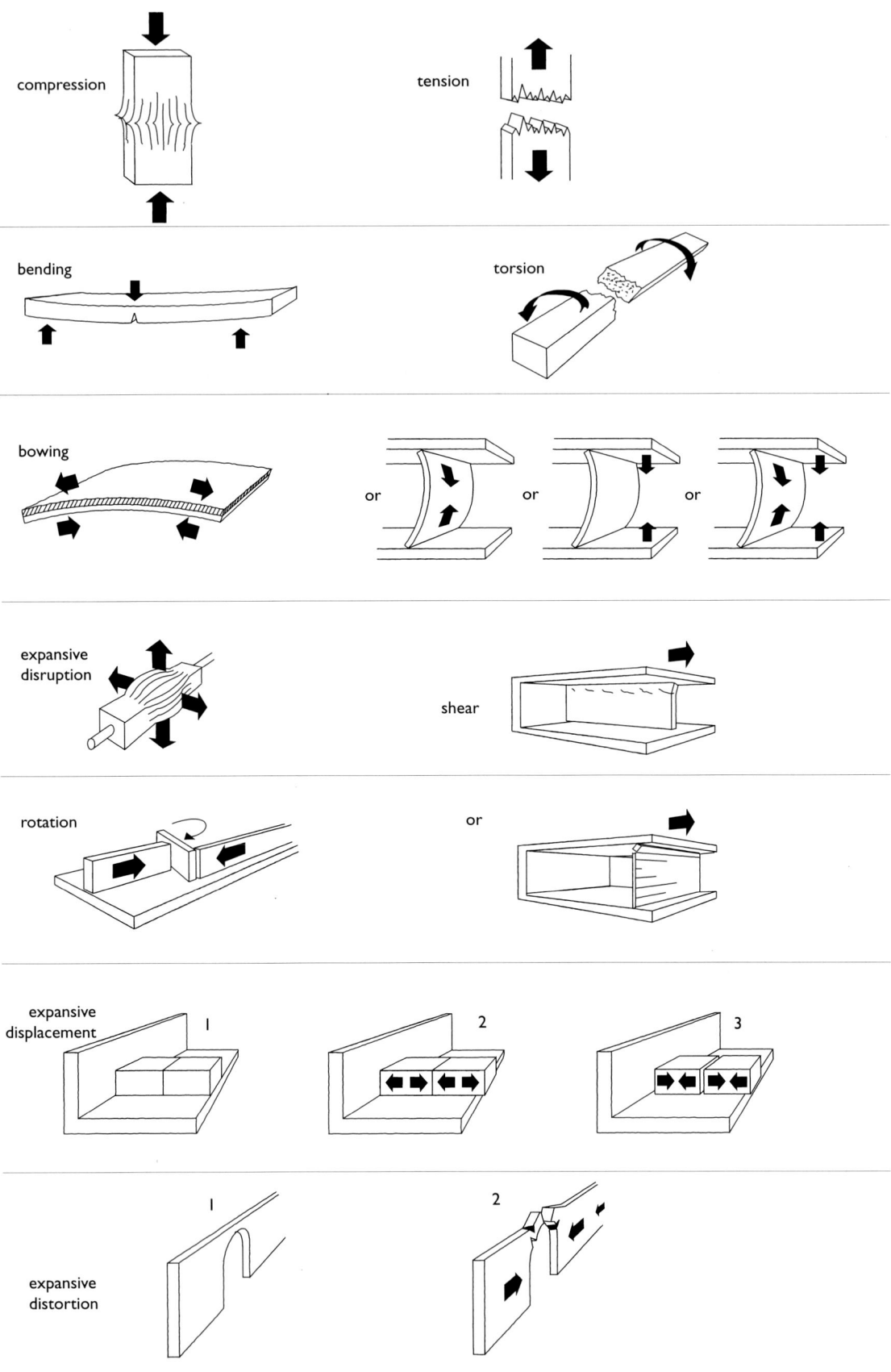

Figure 2.1
Examples of cracking mechanisms

TABLE 2.1 MODULUS OF ELASTICITY VALUES

Note: unless more specific data are available, design should be based on the higher value where a range is shown

Material	Modulus of elasticity E (kN/mm^2)
Natural stones	
Granite	20–60
Limestone	10–80
Marble	35
Sandstone	3–80
Slate	10–35
Cement based composites	
Mortar and fine concrete	20–35
Dense aggregate concrete:	
gravel aggregate	15–36
crushed rock (except limestone)	15–36
limestone	20–36
Steel fibre reinforced concrete	20–41
Aerated concrete	1.4–3.2
Lightweight aggregate concrete:	
medium lightweight	8
ultra lightweight (exfoliated vermiculite and expanded perlite)	†
Asbestos cement	14–26
Glass reinforced cement	20–34
Calcium silicate based composites	
Asbestos wallboard and substitutes	8–10
Asbestos insulating board and substitutes	2.6–3.6
Gypsum and gypsum based composites	
Dense plasters; plasterboard	16
Sanded plasters	8.5–16
Lightweight plasters	1.5–4
Glass reinforced gypsum	16–20
Brickwork, blockwork and tiling	
Concrete brickwork and blockwork:	
dense aggregate	10–25
lightweight aggregate (autoclaved)	4–16
aerated (autoclaved)	3–8
Calcium silicate brickwork	14–18
Clay or shale brickwork or blockwork	4–26
Clay tiling	†
Metals	
Cast iron	80–120
Plain carbon steel	210
Stainless steel:	
austenitic	200
ferritic	200
Aluminium and alloys	70
Copper	95–130
Bronze	100
Aluminium bronze	120
Brass	100
Zinc	140 parallel to rolling / 220 perpendicular to rolling
Lead	14

Material	Modulus of elasticity E (kN/mm^2)
Wood and wood laminates*	
Softwoods	5.5–12.5 (1)
Hardwoods	7–21 (1)
Plywood	6–12 (1)
Blockboard and laminboard	7–11 with core (1) / 5–8 across core (1)
Woodchip and fibrous materials	
Hardboard	3.0–6.0
Medium board	1.7–3.3
Softboard	†
Chipboard	2.0–2.8
Wood-wool cement	0.6–0.7
Rubbers and plastics, etc	
Asphalt	†
Pitch fibre	†
Ebonite	†
Thermoplastics:	
PVC, PVC-U and PVC-C	2.1–3.5
polyethylene (low density)	0.1–0.25
polyethylene (high density)	0.5–1.0
Polypropylene	0.9–1.6
Polycarbonate	2.2–2.5
Polystyrene	1.7–3.1
Acrylic	2.5–3.3
Acetal	2.8–3.7
Polyamide	1.0–2.7
ABS	0.9–2.8
Thermosets (laminates):	
phenol and melamine formaldehyde	5.5–8.5
urea formaldehyde	10
Cellular (expanded):	
PVC	†
phenolic	†
urea formaldehyde	†
polyurethane	†
polystyrene	†
Reinforced:	
GRP (chopped strand)	6–12
carbon fibre (orientated)	180–220
Glass	
Plain, tinted and opaque	70
Foamed (cellular)	5–8

(1) At 12% moisture content; values reduce at higher moisture contents
† No data available

CRACKING IN BUILDINGS The mechanism of cracking

expected size change in service. Surveyors and building failure investigators may need to work back from the amount of bow that has occurred to the corresponding linear size change as part of the evidence as to causes.

A sufficiently good approximation, for these purposes, of the amount of bow for a given linear size change in a thin, unrestrained component is easily obtained if bending resistance is ignored. Thus, in Figure 2.2, if the arc length s is taken as the unrestrained expanded length and c as the original length and b_m is the amount of bow, then:

$$\text{Strain} = \frac{(s-c)}{c}$$

But, as noted earlier, for temperature-induced size changes:

$$\text{Strain} = \alpha\, t$$

The nomogram in Figure 2.3 displays the relationship between bow (expressed as a proportion of the chord length) and strain. An example of its use in a calculation is given in the feature panel alongside.

Two important conclusions result:

- designers need to make provision for linear size change if comparatively large distortions are to be avoided
- surveyors should not assume that comparatively large distortions necessarily imply that significant size changes or displacements have occurred in the structure

Also while, in the above example of bow, expansion of the contained member was assumed, it could equally have been assumed that contraction of the containing structure (or both expansion and contraction simultaneously) was occurring.

AN EXAMPLE OF CALCULATION OF BOW DUE TO TEMPERATURE CHANGE

Suppose that a 2 m (2000 mm) long thin section of material with a thermal coefficient α of 25×10^{-6} is subjected to a temperature change, after fixing in position, of +50 °C:

Strain = $\alpha\, t$
= $25 \times 10^{-6} \times 50$
= 0.00125 (or 1.25×10^{-3})

From Figure 2.3 the corresponding value of b_m/c is 0.021. Thus bow:

$b_m = 0.021 \times c$
= 0.021×2000
= 42 mm

Since bending resistance has been ignored, the true amount of bow will be slightly less. However, it is important to note that in principle the amount of bow is very considerably greater than the linear size change that promoted it. (In the above example the linear size change assumed was 2.5 mm, which produced a calculated bow of 42 mm.)

The mechanisms operating to produce cracks or distortion in a building are complex and (to the extent that buildings differ markedly in size, shape, the materials used, the conditions experienced, the loads carried and the structural form) each failure mechanism is peculiar to the circumstances of that particular building. Identification of the mechanism that will operate is essential if design is to be such that cracks are minimised and if diagnosis of the cause of a crack is to be well founded.

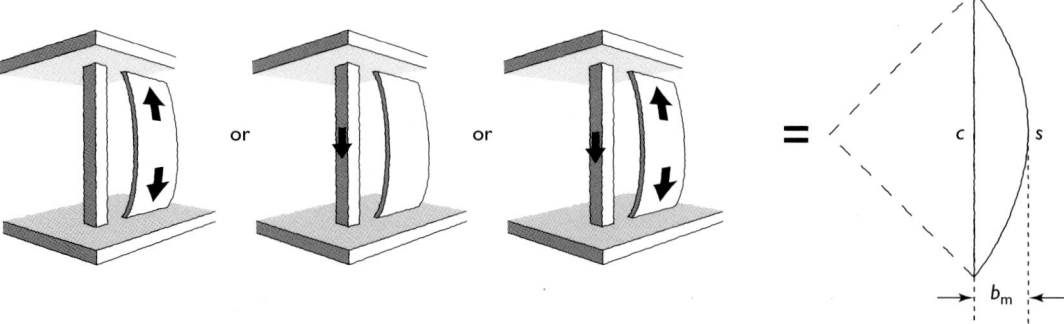

Figure 2.2
Demonstration of bow

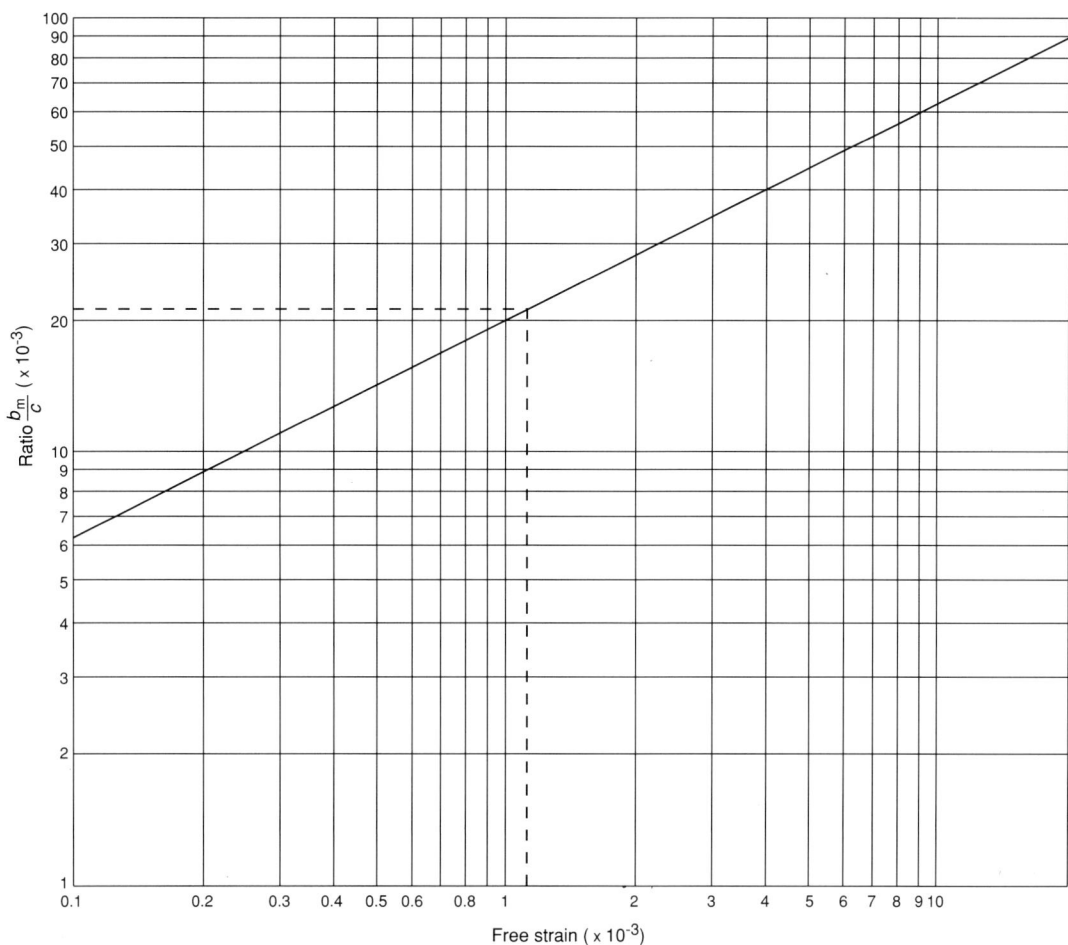

Figure 2.3
Relationship of bow to arc length and chord length (approximate estimation of bowing)

Chapter 3
Joints as safeguards against cracking

Most of the cracking that arises in a building has its origin in daily and seasonal size changes, phenomena which are both certain and continuous throughout the building's life. But the extent and distribution of such cracking depends on the provisions made to accommodate size changes. For this reason joints have a crucial part to play. Careful design of joints can certainly minimise and often eliminate cracking due to size changes. In this context, joints are of two types:

- those which are introduced expressly to accommodate size changes – in other words, movement joints
- those which exist because separate units are assembled to form the fabric of the building – assembly joints

In the second type, three points must be kept in mind.

- Such joints will usually be subjected to the effects of size changes, both in the units and in the parts of the building supporting them.
- The number of joints amongst which size changes are shared, and the relative strength of the jointing materials and the units, will influence the distribution and sizes of cracks.
- An assembly of jointed units may behave as a larger unit so that it is the joints at the perimeter of the assembly that largely determine crack sizes and distributions.

This chapter deals, first, with movement joints provided expressly to accommodate size changes, and then with the more complex considerations underlying provision for size changes at assembly joints.

Movement joints

Movement joints are often called expansion joints since their primary purpose is to provide for expansive size changes and so avoid compressive spalling. Size changes are often reversible (as indicated on page 5). Even when a particular size change is indeed irreversible, it is virtually certain that other reversible size changes will be superimposed on it. In practice, therefore, so-called expansion joints must accommodate both expansion and contraction. Joints of the same kind may also be needed to accommodate contraction of the containing construction. When put to that use the joints are often referred to as soft joints.

Thus the primary purpose of movement joints is to accommodate, at specific points, dimensional changes in the construction. The origin of those changes may sometimes be near to or remote from the joint, as some of the mechanisms illustrated in Figure 2.1 indicate.

Most often movement joints are formed by incorporating, in the course of construction, a joint filler which acts as a spacer and so can ensure that the joint width is adequate. Clearly the compressive strength of such fillers must be less than that of the adjoining construction if the latter is not to be damaged by forces acting to close the joint. The filler must also have an adequate recovery to deal with size changes that subsequently open the joint. Table 3.1 (on page 20), taken from BS 6093[5], provides data for some typical filler materials. An alternative approach, which may sometimes be possible, is to use a filler strip to control the forming of the joint to an adequate width but then to remove the strip and substitute a material with better movement accommodation characteristics – precompressed impregnated plastics foam strip, for example.

In principle a joint which has no other function than to permit size changes (as might be true of a movement joint in a freestanding wall, for example) does not need to be filled or sealed. In practice, however, it will be important that detritus cannot block the joint and prevent its intended operation. In most cases the joint will be required to perform secondary functions also, such as the exclusion of rainwater. These considerations dictate the need for a seal of some kind. If sealants are to be used for this purpose, regard must be given to the requirements of the sealants themselves; these requirements will apply to both movement joints and assembly joints at which movement will occur.

TABLE 3.1 FILLERS FOR MOVEMENT JOINTS

Joint filler type	Property					
	Typical uses	Form	Density range (kg/m³)	Pressure for 50% compression (N/mm²)	Resilience (recovery after compression) (%)	Tolerance to water immersion
Wood fibre/bitumen	General purpose expansion joints	Sheet, strip	200–400	0.7–5.2	70–85	Suitable if infrequent
Bitumen/cork	General purpose expansion joints	Sheet	500–600	0.7–5.2	70–80	Suitable
Cork/resin	Expansion joints in water-retaining structures where bitumen is not acceptable	Sheet, strip	200–300	0.5–3.4	85–95	Suitable
Cellular plastics and rubbers	Low-load transfer joints	Sheet, strip	40–60	0.07–0.34	85–95	Suitable if infrequent
Mineral or ceramic fibres and intumescent strips	Fire-resistant joints: low movement	Loose fibre or braided and strip	Dependent upon degree of compaction	Dependent upon degree of compaction	Slight	Not suitable

Sealants can be classified broadly as:

- plastic
- plasto-elastic
- elasto-plastic
- elastic

The selection of a sealant of the appropriate class is influenced by the expected rate of change of size to be accommodated. The rate of change will depend on factors such as the thermal inertia of the construction. For instance, lightweight components of metal or plastics respond rapidly to the ambient temparature changes, while other dimensional changes may be spread over long periods – creep in structural members, for example. Broadly, size changes that occur slowly will permit the use of sealants towards the plastic end of the range of classes, but more rapid size changes will dictate the use of sealants from the elastic end of the range.

The width-to-depth ratio of the installed sealant has a significant influence on it's performance and durability in service. The requirements of each class are given in Table 3.2.

Further to these ratios, in narrow joints there should be a minimum sealant depth of 10 mm when jointing surfaces are porous and 6 mm when non-porous. These apply to all classes of sealants (BS 6213[6]).

A crucially important characteristic of elastic and elasto-plastic sealants is their Movement Accommodation Factor, or MAF. The MAF of a sealant quantifies its ability to accommodate tensile strain. Irreversible size changes that reduce joint width should therefore be allowed for additionally. The following expression allows calculation of the minimum joint width Ws_{min} determined by MAF alone:

$$Ws_{min} = \frac{TRM \times 100}{MAF} + TRM$$

where TRM is Total Relevant Movement (BS 6093).

TABLE 3.2 WIDTH-TO-DEPTH RATIOS OF SEALANTS

	Width	Depth
Plastic	1	1 to 3
Plasto-elastic	1	1 to 2
Elasto-plastic	2 to 1	1
Elastic	2	1

In this form the expression ensures that, after any expected irreversible closing movements have been added to Ws_{min} (and the effects of inaccuracies taken into account, described later on this page), the MAF will not be exceeded in service.

The provision of adequate joints is important to the avoidance of cracking in the fabric of the building. The correct design of sealant filled joints is also important to the avoidance of cracking in sealants, whether in the body of the material or at its interface with adjoining construction.

The remainder of this chapter has only limited relevance to the design of (or the diagnosis of failures in) buildings in which the main fabric is wholly built of masonry units. In many kinds of building, however, including those with substantial amounts of masonry in their construction, the building sequence dictates that some parts have to be fitted into spaces formed by parts already in place. There will be many joints in the construction which arise solely from this circumstance. It is convenient to describe them as assembly joints.

Assembly joints

Assembly joints, particularly in anything other than masonry, are often seen as inconvenient but unavoidable gaps in the construction that have to be filled. The reality is that, unless provision is made at some other point, they must accommodate the size changes that inevitably occur or cracks will almost certainly result somewhere in the construction. The process by which appropriate provision can be made in design – taking account where necessary of MAFs – is the same for assembly joints as for movement joints but with one critically important difference.

It has been supposed, when describing movement joints, that a joint of that kind is constructed to a predetermined width – for example by the use, as is common practice, of a strip of material acting as a spacer. Only very exceptionally are assembly joints put together in this way. An intended joint width may be specified, but the width achieved is dependent on the accuracy of size and position of the assembled units. In other words, any dimensional provision made to accommodate size changes at assembly joints is potentially overturned by inaccuracies in components and in site construction. At this stage, therefore, it is necessary to consider the fundamental character of inaccuracy.

Inaccuracies in building

In the field of building design and construction the subject of accuracy and tolerances is not generally well understood, and this is particularly true of the terminology. The effect is that design in particular is less competent in this respect than it could be. Perhaps the commonest misconception is that tolerances are a means to describe how accurate (or inaccurate) something is. However, the nature of inaccuracy is such that it cannot be so described, as is now explained.

Suppose that a fairly large number of pieces of wood 300 mm long are needed, to be sawn by hand from long lengths. Using a rule and a pencil each 300 mm length is roughly marked out and sawn off in turn. Suppose that all the cut pieces are then very carefully measured and the sizes so measured are plotted against the number of pieces. The graph produced will have the characteristic shape shown in Figure 3.1 and is known as normal distribution.

Figure 3.1 shows that the largest single group of sizes produced is that at 300 mm – the size that the pieces were indeed intended to be. However, if the pieces that proved to be 299 mm long are added to those that proved to be 301 mm long, these pieces together outnumber those that proved to be 300 mm long. If all the pieces that proved not to be 300 mm long are added together they greatly outnumber those which were 300 mm long. This characteristic is common to all processes subject to random error; it is no one's fault that the pieces of the required size are outnumbered by the remainder, or that some sizes produced differed

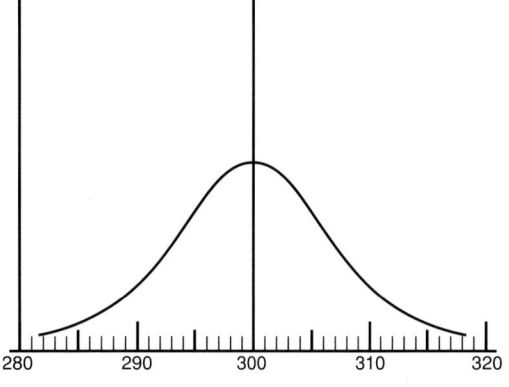

Figure 3.1
Normal distribution without using a jig

greatly from the intended size. The number that differed greatly from 300 mm was small – or, to put it another way, their probability of occurrence was low. But when they occur they cannot be held up as examples of shoddy workmanship: their occurrence, at that level of probability, was inevitable when that particular process was used to manufacture those particular items.

Suppose now that in that same process the hand saw was located in position by some kind of jig or template. The distribution of sizes now produced might be that in Figure 3.2.

The graph of numbers (or frequency) against sizes still has the characteristic normal distribution shape but this time there are more pieces matching the required 300 mm. While they are still outnumbered by the sizes that are not 300 mm, sizes which differ greatly from 300 mm are still produced, but the probability of their occurrence has been reduced.

This increased probability of occurrence of the intended 300 mm size, and the reduced probability of unwanted sizes, together mean that the new process is intrinsically more accurate. Similarly, for every kind of component or typical piece of site construction, its dimensional accuracy is dependant on both what it is and how it is made, and that accuracy is characteristic of all such items made in that way.

If Figure 3.2 is now superimposed on Figure 3.1 and two limits of size are marked out on the base line at, say, 15 mm each side of the centre, it will be seen (Figure 3.3) that both sets of timber pieces could be described as complying with a tolerance of plus or minus 15 mm, but the truth is that the accuracies of the two processes are very different. This is one important reason why the term tolerance should not be used when a means of describing accuracy is sought.

A further point must now be considered. When sawing off the 300 mm lengths of wood, it is quite likely that the saw would consistently be positioned to one side or other of the marked line. For this reason the plotted graph in Figure 3.1 might well be found to be centred on, say, 298 mm rather than the intended size of 300 mm and all of the measured sizes in the curve correspondingly displaced by 2 mm from their positions in Figure 3.1. This kind of circumstance is called, for fairly obvious reasons, systematic error and the distribution curve would be said to have a mean, in this example of –2 mm.

It can now be seen that there are two characteristics that describe accuracy: the shape of the distribution and the mean. Comparing accuracies (eg, of the same product produced by several different manufacturers) by comparing the shapes of their distribution curves is clearly not a simple process. What is needed therefore is a single numerical value that describes the shape of an entire distribution, so that these single numerical values can be compared.

Suppose that one of the pieces of sawn timber is selected at random from those from which Figure 3.1 was plotted. It proves to have a length of, say, 305 mm and thus it could be described as differing from the mean by 5 mm. It follows, then, that the single numerical value that describes the entire distribution curve is the average amount by which all the pieces differ from the mean. It can be shown as a mathematical expression: if the mean is

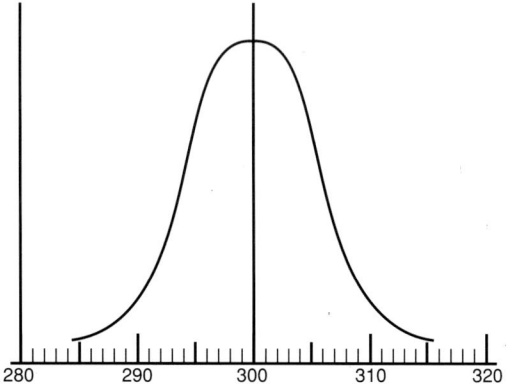

Figure 3.2
Normal distribution using a jig

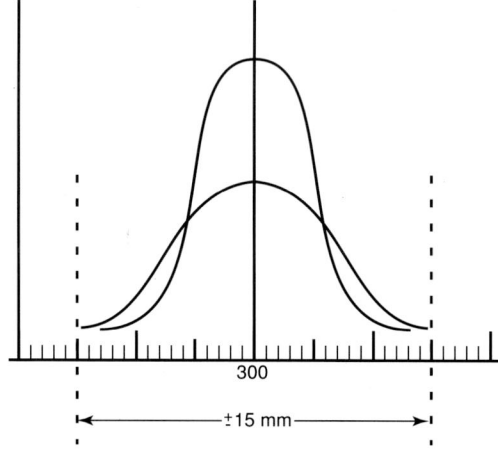

Figure 3.3
Normal distribution: using and not using a jig compared

designated by $\bar{x}$ and an individual size by x_i, then the difference between any individual size and the mean is $(\bar{x} - x_i)$. To obtain the average of all the differences then – as for any average – they must be summed (the mathematical symbol is Σ) and divided by the number of them n. Thus:

$$\frac{\Sigma(\bar{x} - x_i)}{n}$$

However, since the curve is symmetrical about the mean, for every positive value of the difference there is an equivalent negative value and the sum would thus be zero. A solution is to square all the values before summing them, so that the negative values become positive. Thus:

$$\frac{\Sigma(\bar{x} - x_i)^2}{n}$$

This mathematical device must now be corrected by taking the square root of the result:

$$\sqrt{\frac{\Sigma(\bar{x} - x_i)^2}{n}}$$

The value so obtained represents the average amount by which all of the values, such as those plotted in Figure 3.1, differ from the mean; it thus encapsulates the entire shape of the distribution curve. This numerical value is called Standard Deviation or s. The greater the amount by which all the individual sizes differ from the mean the larger the calculated value for s will be.

If the batch of pieces of sawn timber used to produce Figure 3.1 is sufficiently large it can be taken to be representative of all such pieces of timber produced in that way. The values found for the mean $\bar{x}$ and the Standard Deviation s would be valid for all, including further batches that will not be sawn until, say, ten years hence. The statistical term for 'all' is the population, an infinite number. However, if infinity ∞ is substituted for n in the expression above, the calculation will produce the result of any finite number divided by infinity: zero. The mathematical device employed to overcome this is to substitute $(n-1)$ for n, thus dividing by one less than infinity. The expression so modified becomes the expression for Standard Deviation of the population and, to distinguish that Standard Deviation from one calculated for the sample or batch, it is designated σ instead of s.

Thus:

$$\sigma = \sqrt{\frac{s(\bar{x} - x_i)^2}{(n-1)}}$$

So if representative samples of parts of building construction, built by normal techniques, are measured, the calculated value for the mean x and the Standard Deviation σ will provide a complete description of the accuracy of all such pieces of construction. Thus they provide a description of the accuracy of such items yet to be made.

A remarkable property

Standard Deviation has a property which is of immense potential value to the process of designing and building.

The value $\pm 3\sigma$ always embraces 99.73% of the distribution, while $\pm 2\sigma$ embraces 95.45% and $\pm \sigma$ embraces 68.27%. Suppose that the value of σ, calculated from measurements of a representative sample such as the pieces of sawn timber whose sizes are plotted in Figure 3.1, is 5 mm. Then ± 15 mm would contain nearly 100% (strictly 99.73%) of the pieces, ± 10 mm would contain 95% and ± 5 mm would contain 68%. These percentages could equally well be expressed as probabilities. For example, since $\pm 2\sigma$ embraces 95%, it follows that 5% would be outside that range; one might say therefore that a piece of sawn timber taken at random would have a 5% chance, and thus a 1 in 20 chance, of being outside the range covered by $\pm 2\sigma$. A standard graph, reproduced in Figure 3.4 (on page 24) and based on that in BS 6954:Part 3[7], provides probabilities for all multiples of σ likely to be needed.

Once a representative sample (of, say, a commonly occurring piece of construction in buildings) has been measured and the value of σ calculated, the graph can be used to find the probability of occurrence for any chosen size from among those scattered around the intended size – and the probability therefore of any size proving to be outside a predetermined acceptable limit even though construction may not start for some years.

Representative samples of very many common pieces of building construction have been measured. The corresponding calculated values of mean x and Standard Deviation σ are given in

CRACKING IN BUILDINGS Joints as safeguards against cracking

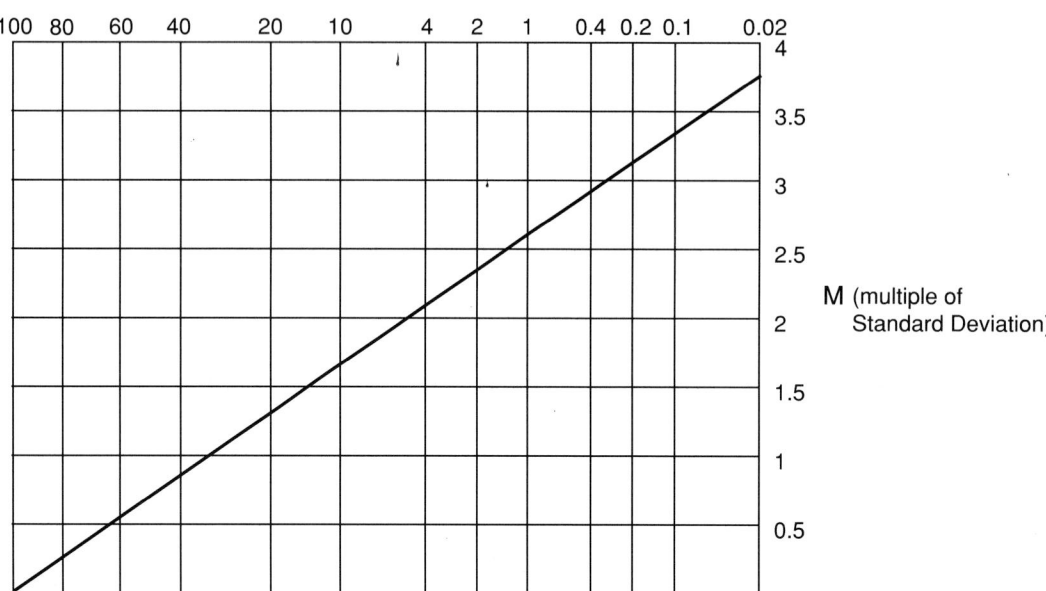

Figure 3.4
Percentages outside chosen limits: the percentages are halved to find the proportion outside a chosen upper or a chosen lower limit

TABLE 3.3 EXAMPLES OF MEAN AND STANDARD DEVIATION FOR PARTS OF BUILDINGS

Spaces between elements	Mean (mm)	Standard Deviation (mm)
Brick walls at floor level	2.0	6.4
Brick walls at soffit level	0.3	9.7
In situ concrete walls at floor level	−1.9	7.3
In situ concrete walls at soffit level	−0.5	8.9
Steel columns at floor level	−1.3	5.5
Steel columns at soffit level	−0.8	4.8

Tables 4 and 5 of BS 5606[8]. Then using the given value of σ for a particular piece of commonly occurring construction to enter the graph in Figure 3.4, the probability of any particular size occurring in the same kind of construction can be known in advance.

Some typical values for σ are taken from Table 4 of BS 5606 and reproduced in Table 3.3 as examples of the order of size of such values. A very simple illustration of their use (leaving aside for this purpose the effect of $\bar{x}$ being something other than zero) is shown in the feature panel in the next column.

Tolerances

In a piece of building construction, as in a piece of machinery, some of the dimensions must lie within certain limits if the parts involved are to fit together and the assembly to work as intended. Those limits are determined by the nature and use of the assembly and it is part of the designer's task to identify, objectively, what they are. The limits, when found for any particular dimension, represent the maximum error permitted by the nature and use of the assembly about the intended or target size for that dimension.

In other words, the limits represent the tolerable amount of error in the size. The term for such limits is tolerance. Thus tolerances describe the limits of tolerable error dictated by the need for the assembly to fit together and to function as intended. It will be clear that to have misused

---

AN EXAMPLE OF PROBABILITY OF OCCURRENCE

Suppose that it is necessary to impose a tolerance of ±12 mm on the verticality of in situ concrete columns of storey (3 m) height. Dividing this tolerance by the value given for σ for that construction (5.7 mm) gives a result of 2.1. Entering the graph in Figure 3.4 at this figure reveals a probability of approximately 4% (or 1 in 25) that such a tolerance will not automatically be met. In other words, there is a 1 in 25 chance that corrective work will be needed.

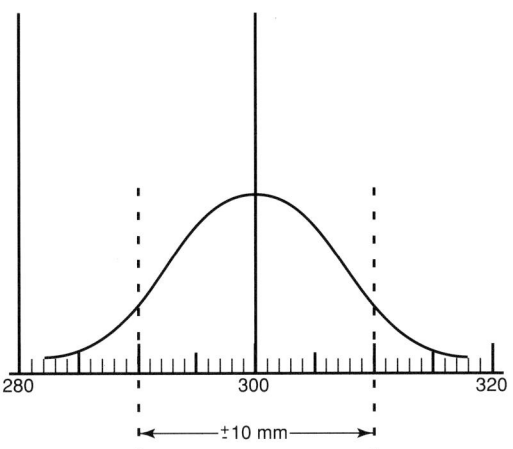

Figure 3.5
Tolerance imposed on normal distribution

the term to describe the accuracy of something (ie, the errors that occur) would have pre-empted the proper use of the term to describe the amount of those errors that is tolerable if the assembly is to fit and function.

The proper use of the term tolerance provides a means to compare what will inevitably happen (as represented in principle by Figure 3.1) with what is required. In Figure 3.5 the distribution curve in Figure 3.1 is repeated with the addition of tolerable limits of error. It is supposed that the use to which the pieces of sawn timber will be put requires that the pieces used in an assembly will have sizes lying within ±10 mm of the intended 300 mm size. If the calculated Standard Deviation σ for all such pieces of timber is, say, 5 mm, then a tolerance of ±10 mm embraces ±2σ. So 95% (strictly 95.45%) of all the pieces produced can be expected to comply while 5%, or 1 in 20, will not.

If the tolerance could be increased (in this example to, say, ±15 mm) then virtually all of the pieces would be within tolerance and usable in the assembly. Such an increased tolerance would be more easily met and would cost less, but the increase cannot be obtained merely by changing the figure. If the tolerance of ±10 mm was correctly identified in the first place, an increased tolerance can be obtained only by changing the design – to one more tolerant of error. An alternative would be to improve the intrinsic accuracy of the pieces of sawn timber. This also cannot be achieved merely by hopeful specification: the process must be changed to one with better characteristic accuracy, as exemplified in Figure 3.6.

Without the information about mean $\bar{x}$ and Standard Deviation σ contained in BS 5606, the need for either a more tolerant design or an intrinsically more accurate process could not be recognised before construction began. It might remain unrecognised until joints, with widths unacceptably reduced by inaccuracy, closed completely in service, and cracks and associated damage developed in the structure.

Joints and accuracy

The foregoing description of the nature of inaccuracy and its relationship to tolerances provides a background against which to deal with the dimensional needs of joints. (For this purpose, joints formed by rigid mechanical connection – bolted, riveted or welded for example – can be discounted: their effect is simply to create a larger unit).

In the context of cracking in buildings, the essential requirement of joints is that they accommodate size changes in the various parts of buildings so that damaging stress in those parts is avoided. Although this requirement might be met in many cases by designing to ensure only that adjoining components do not expand into edge-to-edge contact (ie, that joint widths are not potentially reduced in service beyond zero), it would be unrealistic to treat joint design in these terms alone: joints inevitably have other needs also.

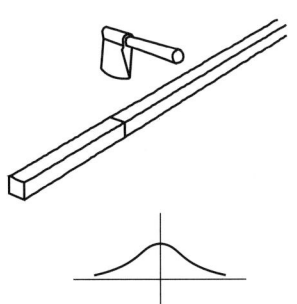
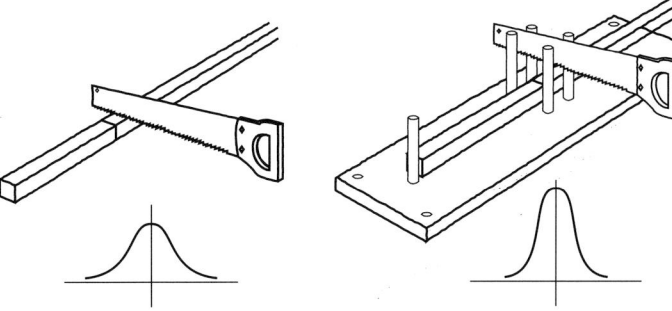

Figure 3.6
Improving characteristic accuracy

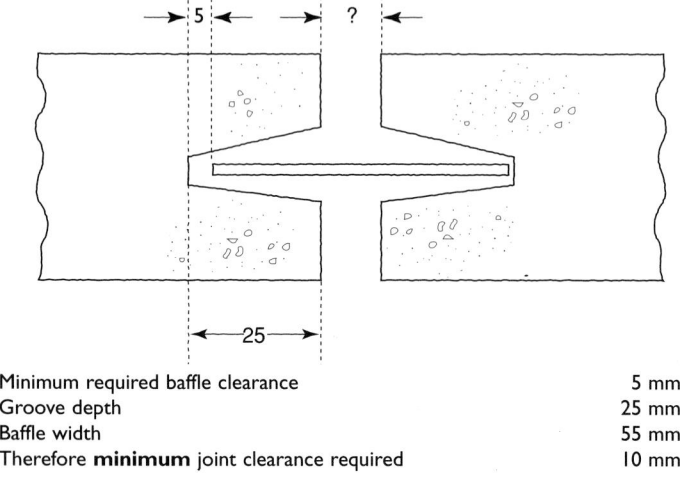

Minimum required baffle clearance	5 mm
Groove depth	25 mm
Baffle width	55 mm
Therefore **minimum** joint clearance required	10 mm

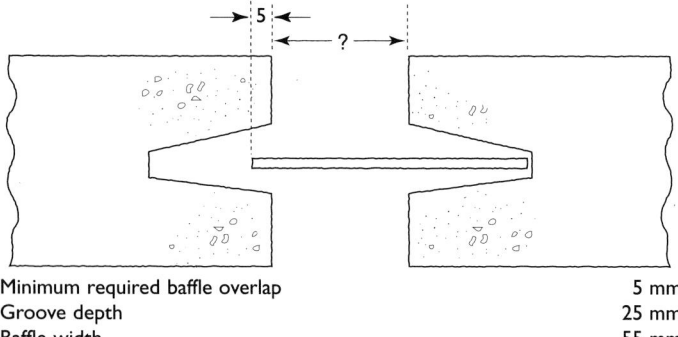

Minimum required baffle overlap	5 mm
Groove depth	25 mm
Baffle width	55 mm
Therefore **maximum** joint clearance required	25 mm

Figure 3.7
An example of examination of joint design for its width (clearance) requirements

The key issue is that each design of joint carries requirements for a minimum and a maximum width (or clearance). This is most readily illustrated by the joint in Figure 3.7 (taken from BS 6093) in which the chosen dimensions for the width of a baffle and for the grooves containing it impose a minimum joint clearance of 10 mm and a maximum of 25 mm.

It will be seen that these requirements impose a tolerance on the normal distribution of inaccuracies at the joint. If the inaccuracies (described by mean and Standard Deviation) are compared with the joint tolerance, it will be possible to specify target sizes for the components so that the probability that assemblies will automatically have joints within the required clearance limits is optimised. Figure 3.8 shows some cases that potentially arise at joints.

In Figure 3.8a, 50% of such assemblies would theoretically produce joint clearances of less than zero. (In practice, of course, they simply could not be assembled.) The other 50% of such assemblies would contain many potential tight fits (joint clearances of zero or only a little more) and there would be high risk of joints closing completely in service, with resulting cracking and damage.

In Figure 3.8b, the chosen target size for the component is such that 50% of assemblies would produce joints with less than the required minimum clearance. The risk of damaging edge-contact is less than that in Figure 3.8a but remains high; a still smaller target size must be specified.

In Figure 3.8c, the chosen target size for the component is small enough to reduce the risk of joint clearances being smaller than the required minimum to an acceptable level.

Finally, in Figure 3.8d, the chosen target size has been further reduced to the point that the risk of the maximum required joint clearance being exceeded is about to become significant. Thus any chosen target size for the component between that in Figure 3.8c and that in Figure 3.8d will carry minimal risk that the joints in any assembly will not be within the clearance limits required. The chosen target size for the component then needs to be adjusted to take account of changes of size in service.

The calculation methods that are the equivalent of the foregoing are set out in BS 6954:Part 3. Using those methods with the data for mean and Standard Deviation contained in BS 5606 enables target sizes for the joined parts to be specified such that the probability that joints will fail to accommodate size changes in service is reduced to a minimum.

Joints and fixings

Joints are effectively the only parts of buildings at which otherwise potentially damaging size changes can be safely accommodated. Joint design must recognise the need:

- to ensure that components cannot expand into damaging edge-contact
- to take account of the dimensional requirements of products used to form joints
- to adjust design (target) sizes so that unavoidable systematic and random errors ($\bar{x}$ and σ) will not negate the provisions made

Some prior thought must be given to the effect of restraint. Restraint can modify both the

Joints as safeguards against cracking   CRACKING IN BUILDINGS

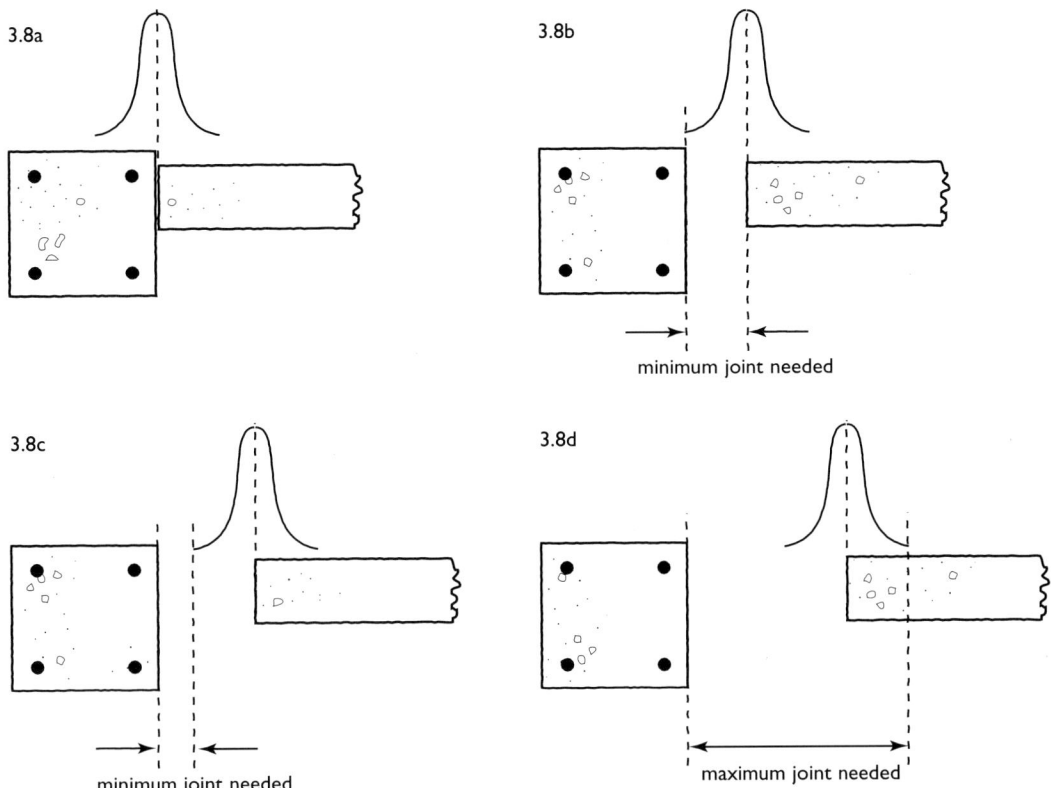

Figure 3.8
The effect of progressively reducing the specified target size

amount of size change and its distribution. Reduction in the amount of size change is probably best discounted in favour of the safe assumption that all of the predicted unrestrained size change will occur. The effect of restraint on the distribution of size changes among joints should also receive some consideration.

In Figure 3.9 (on page 28), two adjacent components each have two fixings at the base into the structure behind. All four fixings are slotted for adjustment at assembly and all are tightened after assembly. Since all four fixings are identical it could be assumed that they all offer equal restraint to size changes.

In practice the chance that restraint offered by each of the four fixings is exactly equal is small. Figure 3.9 illustrates three cases that might therefore arise:

- the joint width is static
- the joint width is reduced when the components expand, but only by the amount of size change in one component
- the joint width is reduced by the total amount of expansion of both components

In a run of such components in, say, a building facade, all three cases may well occur simultaneously. The safe assumption is that the worst condition, shown in Figure 3.9c, will occur throughout the run.

In many items of construction it is readily possible to predict exactly what will be the effect of restraints on the distribution of size changes.

In Figure 3.10 (on page 28), for example, a storey height panel of brickwork is contained between two floors of an in situ reinforced concrete-framed building. It is certain that after construction the columns will shrink to a reduced height and it is very possible that the brickwork will expand as a result of some residual initial moisture expansion in the bricks. Restraint in this case arises very obviously from the entrapment of the brickwork between the floors.

It is equally obvious, therefore, that if a compressible joint is provided at the top of the brick panel, the size changes arising from both causes will occur at that joint and its width will be reduced. The effect of restraint in this instance is to concentrate size changes at that

CRACKING IN BUILDINGS Joints as safeguards against cracking

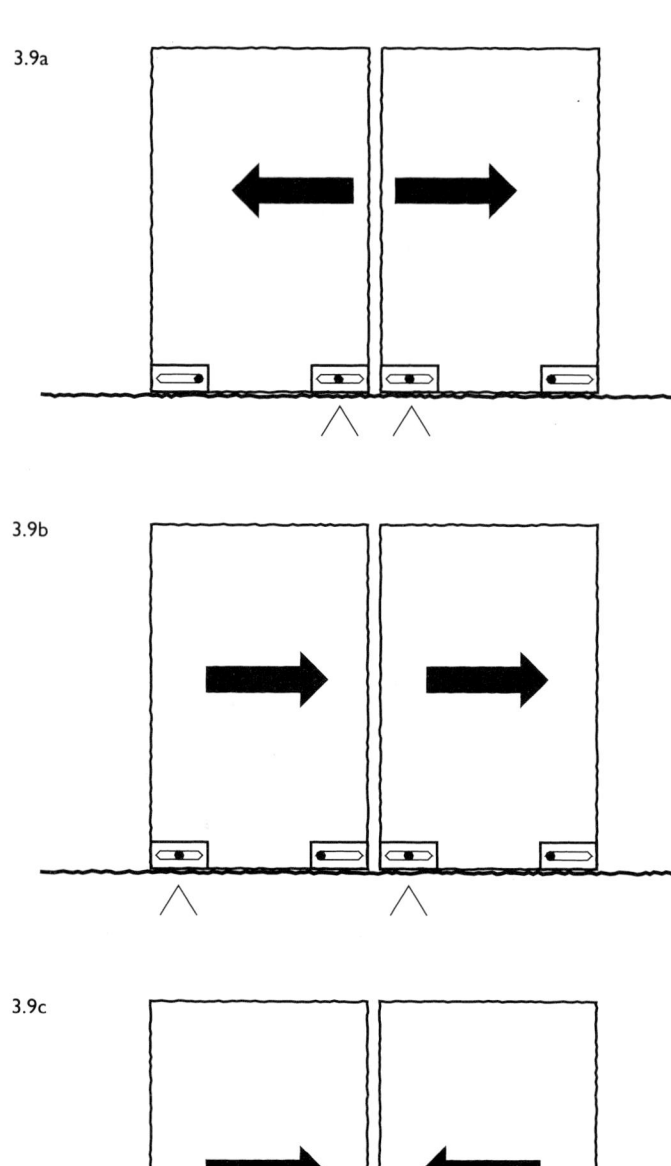

Figure 3.9
Fixings influence size change distribution at joints

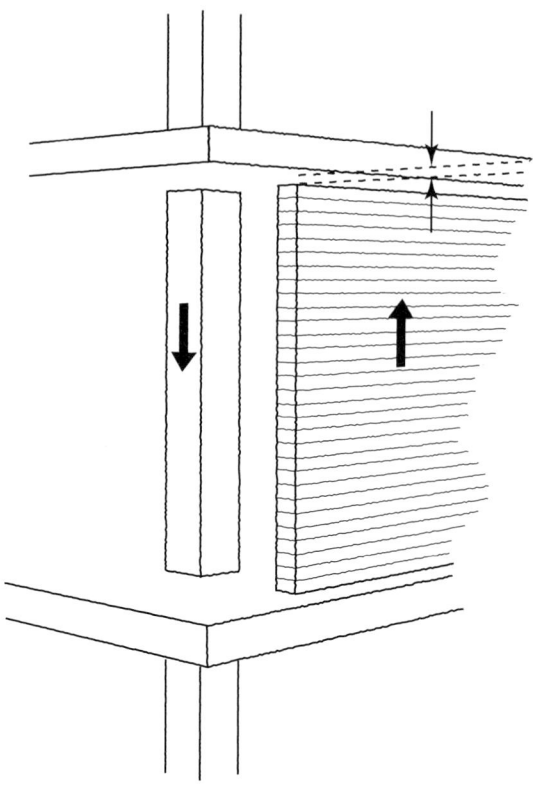

Figure 3.10
Predictable occurrence of size changes

joint and to require the joint, first, to accommodate that reduction in width without risk of complete closure; and, second, to accommodate, when reduced to its minimum by these causes, the effects of other causes – temperature changes, for example. Also, if a sealant is incorporated in the joint, then the sealant's own requirements for joint width must be considered, as explained above.

In other cases, however, the effects of restraints on size change distribution will be less evident; the restraining effects of friction and adhesion, for example, are particularly difficult to quantify. The safest course is to assume that the worst likely condition will occur: some over-provision may result at joints but the likelihood of cracking and similar damage will have been minimised.

The provision made at joints to accommodate size changes will always be a major part of design strategy to minimise cracking; but, where component parts of the building are supported by mechanical fixings of some kind, it may be necessary to make explicit provision in design for the effects of size changes in the components on the fixings themselves.

Two factors are important:

- fixings can be positioned, by design, so as to ensure that size changes accumulate at particular joints where provision is made for them
- slots or similar devices, incorporated in fixings to accommodate size changes in components and structure, will need to accommodate a range of size changes in the same way as joints

These slots will similarly be subject to the effects

of inaccuracies; calculation of the total slot length (or other provision) needed is closely analogous to the calculation of required joint width, omitting that part which takes account of MAFs for jointing products.

Given that joints and fixings are designed in accordance with the principles set out in this chapter, the risk of cracking, spalling and other damage arising from normal daily or seasonal size changes will be minimised. When surveyors or failure investigators set out to diagnose the cause of a crack they will need to consider, in the light of the contents of this chapter, whether the design provision made at joints and fixings to accommodate size changes has been adequate.

Part II
APPLYING THE SCIENCE

Introduction to Part II

Applying the data in Part I to real circumstances in building requires judgement. The conditions experienced by parts of buildings in service are rarely, if ever, exactly known. Assumptions must be made about probable ranges of conditions and responses to them. Also, design decisions about provisions made to minimise cracking may be tempered by considerations of their cost or by the overriding claims of other performance requirements.

The aim of Part II is therefore to show how the science can be applied. How it should be applied in particular cases is a matter for designers to decide in the context of the design of buildings. However, where generally adopted measures exist they are described and are likely to be widely applicable to buildings in the UK climate.

It is not intended that the emphasis should be on design, but design considerations are dealt with first in each section since the presence or absence of features that influence cracking are determined, initially at least, by design decisions. Subsequently those responsible for surveying, investigating failures or maintaining the building, will need to know what features might not have been incorporated in design and, therefore, in the light of the building's performance, which of them, if included, might have reduced or avoided damage.

The earlier editions of well known standard textbooks of building construction are useful sources of information about earlier practices when there is uncertainty about what may be concealed within the existing construction.

It has been noted in Part I that cracks in buildings are only rarely of structural significance. Often they do not materially affect other aspects of performance. Because of the wide variety in the character of buildings and in their usage, a universally applicable list of cases in which cracks significantly affect performance is not possible. However, it is useful to review some instances in which cracks, whether or not structurally significant, may have a marked effect on the performance of other parts of buildings.

Examples are:

- cracks through separating walls and separating floors which can greatly increase airborne sound transmission, even if only a few hundredths of a millimetre wide
- cracks in external rendering on clay brick walls which, if more than hairline, can initiate damaging sulphate attack
- cracks that penetrate impervious coverings on flat roofs which, even if very fine, can be a major source of rain penetration, particularly where roofs are not well drained
- cracks in impervious wall claddings which, because the surfaces are impervious, can be subjected both to high water loads and to large air pressure differentials leading to copious rain penetration
- cracks in hygienic surfaces which, even if very fine, can harbour bacteria
- cracks in tanking membranes which can admit copious amounts of groundwater under pressure
- cracks in vapour control layers which can induce progressive accumulation of condensate in unventilated cavities, usually with significant consequential damage
- cracks in ground floor slabs which, in normal circumstances would be negligible, may be significant where the design is intended to resist the upward passage of gases, including radon and methane, from the ground or from fill materials
- fine cracks in the walls of steel-framed structures which may be the first indication of significant corrosion of hidden structural members

# CRACKING IN BUILDINGS

Chapter 4
Temperature-induced size changes

Perhaps the most important distinguishing characteristic of temperature-induced size changes is that they occur continuously in all buildings, since no building material is unaffected dimensionally by temperature change. In principle the amount of size change is dependent on the range of temperature experienced, the length of the affected part and the thermal coefficient of the material. In practice the amount of change is often reduced by the restraints offered by mechanical fixings, adhesion and friction. Although stresses arising from such restraints may need to be considered in design (Chapters 1 and 2), the safest design provision to accommodate size change is that which will accommodate the full theoretical change. Some reduction due to restraint can be legitimately assumed, but the difficulty of quantifying restraints inevitably introduces a risk of under-provision. The effect of restraint on the distribution of size changes among joints is rather more predictable and should always be considered.

The first step, therefore, is to make an estimate of the temperature range that might be experienced in service. The range will depend on:

- the environmental temperature change
- the colour, exposure and orientation of the parts
- the duration of environmental temperature change in relation to thermal response times of the parts
- the response times as determined by:

  - the mass of the parts
  - insulation protecting the parts from environmental change
  - insulation preventing loss (from the affected parts) of heat gained from environmental change

None of these is precisely quantifiable and their combined effect still less so. The safest course is to identify the worst likely circumstance and the correspondingly greatest size change; then to consider the practical consequences if full provision is not made. For the surveyor this provides a means to assess whether temperature size change alone, even on the worst assumptions, could be responsible for the observed cracking. For the designer it provides a means to assess the risk (which it may be justifiable to take) attached to assumptions about the way the full theoretical size change will be reduced by some of the factors listed above.

The considerations dealt with in this chapter relate to provision for temperature-induced size changes only. Design must ultimately take account of other causes of size change and of building movements. Particular note should be taken that the irreversible expansion of a *contained* structure and the irreversible contraction of a *containing* structure can markedly reduce the clearances provided for temperature-induced size changes.

Walls and cladding
Design principles
For temperature-induced size changes there are broadly two cases to consider.

In Case 1, the wall is long enough to warrant the inclusion of intermediate joints in order to limit the change of size to something less than the cumulative total (Figure 4.1a on page 34).

In Case 2, the size of the wall is not so great as to warrant subdividing it. Design can seek to accommodate the cumulative total size change at the perimeter, where joints may be dictated in any event by the change of construction (as, for example, where infill panels are bounded by a structural frame) (Figure 4.1b).

Thus, in Case 1, a decision has to be made about the lengths into which the wall must be subdivided while, in Case 2, a decision must be made about the amount of size change that will accrue at the perimeter. In both cases the expression:

$$R = \alpha L t$$

applies, but, as will be clear later, in Case 1 it is solved for L and in Case 2 it is solved for R.

CRACKING IN BUILDINGS Temperature-induced size changes

In both cases the temperature prevailing at the time of construction can significantly change the balance between the positive and negative parts of size changes occurring at the joints. For sealant filled joints the formula relating required-joint-width to Movement Accommodation Factor (page 20) takes this effect into account. For other joints, although the conditions that will prevail cannot be known at the time of design, some thought should be given to the possibility that they may be at either extreme of temperature range at the time of construction. For instance, dry construction could proceed even at temperatures well below freezing; but joint widths that would have been adequate if constructed at normal temperatures may not accommodate size changes which, thereafter, will be wholly in the direction which closes the joints.

In both Case 1 and Case 2 there are two principal mechanisms of thermal cracking that potentially arise:

- compression fracture (when insufficient provision has been made for expansive size changes)
- tension fracture (when expansive size change has occurred and subsequent contraction is restrained)

but other possible effects such as displacement, shear, rotation and bowing need also to be identified.

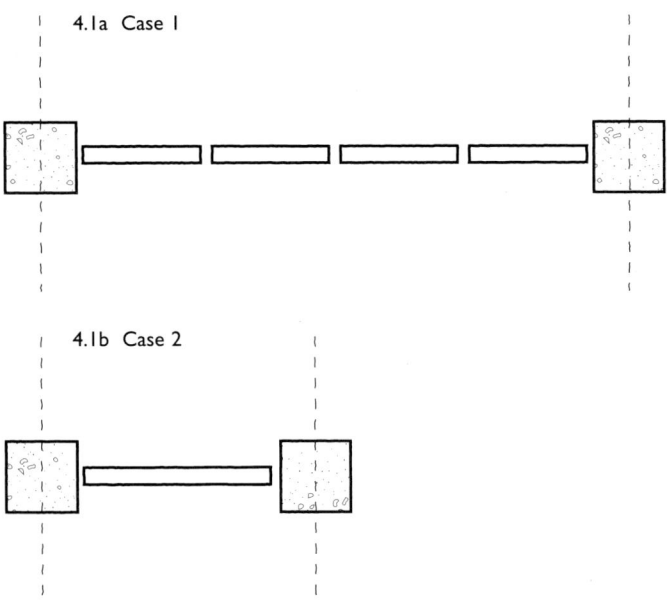

Figure 4.1
Two cases for the design of temperature-induced size changes of a wall: 4.1a, a run of like components, and 4.1b, a single component

Case 1
Where temperature ranges experienced will not be small enough to discount, design freedom in providing for temperature size change is largely limited to the choice of materials (in terms of their thermal coefficients) and the limiting of sizes by introducing discontinuities. Thus, once the material has been selected – or the choice imposed by other considerations – the ability of the joints (at the discontinuities) to accommodate size change is a major determinant of the way in which overall sizes need to be broken down.

The first step is to select one or more joint designs with suitable performance (eg, weathertightness) and compatibility with the materials of the wall. Its capability to accommodate size change must then be assessed, so providing an upper limit for R in the equation:

$$R = \alpha L t$$

which can now be solved for L to provide a corresponding upper limit for size.

Care must be taken to assess how size change will be distributed among the joints as a result of the restraints provided by fixings. Consideration also needs to be given to other possible consequences of size change. For example, the end of a wall bonded at right angles into the wall under consideration might be rotated on plan unless the latter wall is broken into sufficiently short lengths. Containment of the wall by less thermally-responsive construction may be a further factor to be considered; or, conversely, the risk that an uncontained wall could oversail unacceptably (eg, on its DPC) may limit lengths between joints.

Additionally, the location of joints should take into account:

- the location of door and window openings (change of section)
- the location of chases cut into the wall (change of section)
- change in height or thickness of the wall (change of section)
- the opportunity to incorporate reinforcement at changes of section
- the location of major movement joints in other parts of the building
- the location of intersecting walls, piers, etc
- the proximity of corners
- the presence of vulnerable short returns on plan

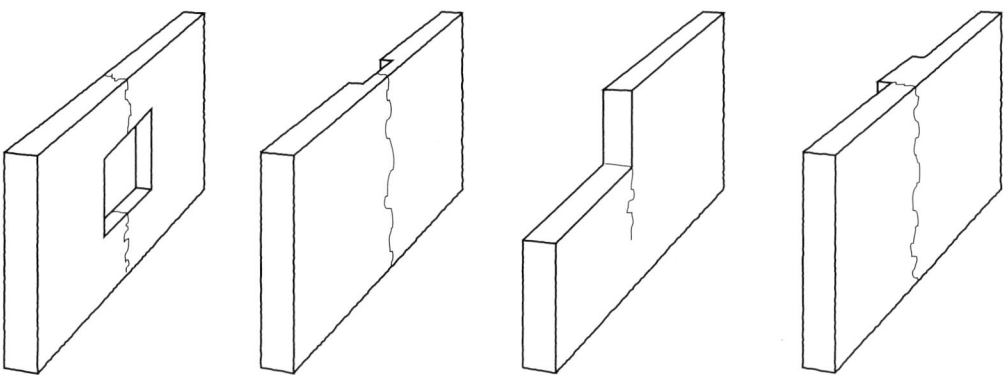

Figure 4.2
Cracks tend to occur at changes of section

Attention is drawn particularly to the items annotated 'change of section' (and to Figure 4.2 above) since it is at these locations that cracks induced during thermal contraction are most likely. The simple rule of thumb to provide joints at all such locations cannot always be applied but should always be considered.

Case 2
In Case 1 the first step was to select a joint with appropriate performance, identify its capability to accommodate size change, and thence deduce the maximum lengths into which the wall should be subdivided. In Case 2 the wall length is small enough not to require subdivision and the first step is therefore to determine the change in length R which the wall will experience for the assumed temperature change. Thus the expression:

$R = \alpha L t$

must be solved for R.

If the wall is contained, for example by a structural frame as in Figure 4.3, then, from joint designs that could provide satisfactory performance, a perimeter joint must be selected that is capable of accommodating its distributed share of R as determined by restraints. Change of size in the horizontal direction often can be safely assumed to be equally divided between the vertical joints at each end, while change of size in the vertical direction will accrue wholly at the upper horizontal joint.

If the wall is not contained then the calculated value of R will indicate whether, for example, distortion at return elevations will be negligible or whether provision to accommodate size changes must be made.

Practical detailing
Clay brick walls
The recommendations of BS 5628:Part 3[9] for unreinforced masonry are not concerned exclusively with temperature size changes. However, the British Standard recommends that vertical joints in unreinforced fired clay brickwork should be positioned not more than 15 m apart to avoid cracking due to thermal contraction. As noted above, where wall lengths need to be subdivided, the joint spacing can be determined by working back from the ability of the chosen joint design to accommodate size changes.

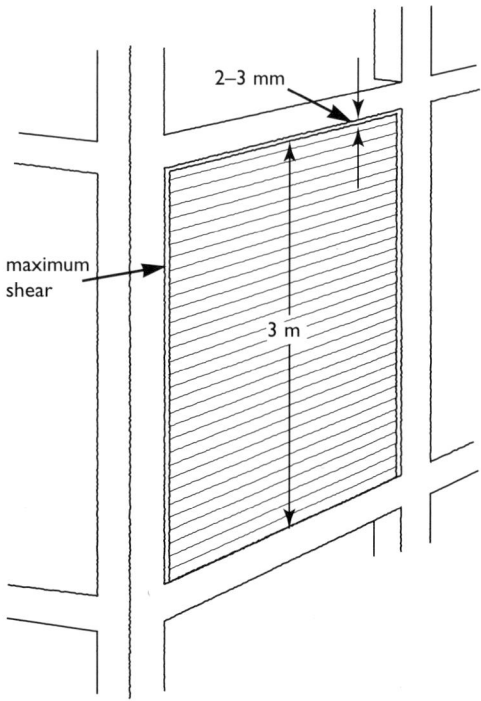

Figure 4.3
A brickwork panel contained by a reinforced concrete frame

CRACKING IN BUILDINGS Temperature-induced size changes

BS 5628:Part 3 gives the useful initial general guide that the width of joints in millimetres should be about 30% more than the joint spacing in metres. Closer spacing than 15 m is likely to be needed where comparatively little restraint is imposed on size changes – as in parapet walls, for example. The vertical joint nearest to a salient or re-entrant corner should be located not further from the corner than half the general joint spacing elsewhere. Additional joints (or, alternatively, reinforcement) may be needed at points of stress concentration – for example at openings which have the effect of introducing changes of section.

BS 5628:Part 3 also draws attention to the possible need to maintain structural continuity at joints (eg, by the use of dowels bridging the joint gap) and the possible need to introduce slip planes to reduce shear stresses between parts that experience different size changes. The standard also suggests that vertical size changes in unrestrained walls are of the same order as horizontal size changes. An example of unrestrained change of size is given in the feature panel below.

AN EXAMPLE OF UNRESTRAINED SIZE CHANGE IN CLAY BRICK WALLS

Using the data in Chapter 2, and assuming a thermal coefficient α of 8×10^{-6} and a temperature range of 85 °C, the unrestrained change in a 15 m length of fired clay brickwork would be in the order of 9–10 mm. The amount of that theoretical size change occurring at any particular joint will depend on the effect of restraint both in reducing the change and in distributing it in particular ways.

Calcium silicate brick walls
BS 5628:Part 3 recommends that, as a general rule, vertical joints should be located at intervals of between 7.5 m and 9 m, and that the ratio of length to height of each panel between joints should not exceed 3:1. As with clay brickwork, it may be necessary to introduce additional joints, or to incorporate reinforcement, at points of stress concentration.

It should be noted that although contraction is the predominant characteristic of calcium silicate bricks, their thermal coefficient is greater than that of fired clay bricks (Table 1.1). An example of unrestrained change of size is given in the feature panel at the top of the next column.

AN EXAMPLE OF UNRESTRAINED SIZE CHANGE IN CALCIUM SILICATE BRICK WALLS

Using the data in Chapter 2, and assuming a thermal coefficient α of 14×10^{-6} and a temperature range of 70 °C, the unrestrained change in a 9 m length of calcium silicate brickwork would be in the order of 9–10 mm. The amount of that theoretical size change occurring at any particular joint will depend on the effect of restraint both in reducing the change and in distributing it in particular ways.

Concrete blockwork walls
Here, too, the requirements for joint spacing can be worked back from the ability of a chosen joint to accommodate size changes. However, BS 5628:Part 3 recommends that, as a general rule, vertical joints should be at intervals of not more than 6 m. The standard also notes that the risk of cracking increases if the length of a panel exceeds twice its height. These recommendations apply in the main to half brick thick walls. Provided suitable precautions are taken, for thicker walls the distances between movement joints may be increased. At changes of section, similar considerations to those noted above for other masonry materials apply. An example of unrestrained change of size is given in the feature panel below.

AN EXAMPLE OF UNRESTRAINED SIZE CHANGE IN CONCRETE BLOCK WALLS

Using the data in Chapter 2, and assuming a thermal coefficient α of 12×10^{-6} and a temperature range of 70 °C, the unrestrained change in a 6 m length of concrete blockwork wall would be in the order of 4–5 mm. The amount of that theoretical size change occurring at any particular joint will depend on the effect of restraint both in reducing the change and in distributing it in particular ways.

Joints in masonry
Chapter 3 dealt with the relevance of MAFs and of width-to-depth ratios in sealant filled joints. But BS 5628:Part 3 notes also that compressibility of the filler material behind the sealant is possibly the most critical factor. A material should be chosen such that a pressure of about 0.1 N/mm² is sufficient to compress it to half its initial thickness. Materials such as

flexible cellular polyethylene, cellular polyurethane and foam rubber are satisfactory, but hemp, fibre-board and cork should be avoided.

Where short returns on plan cannot be avoided, the risk of cracking resulting from rotation can be reduced by incorporating a shear joint (Figure 4.4). It is unlikely that a similar joint will be necessary in an inner leaf since its temperature in service is not significantly affected by external conditions. Stability of the outer leaf, reduced by the introduction of the joint, can be restored by using cavity ties at closer spacing. Where dimension 'X' (in Figure 4.4) is equivalent to four or more bricks, the risk of visually unacceptable cracking is minimised and the shear joint can be omitted.

In fired clay brick walls, it is recommended that vertical joints are incorporated at 15 m intervals. In parapet walls, where much less restraint arises, BS 5628:Part 3 recommends halving this interval. Similar considerations arise for freestanding boundary walls. For both parapet and boundary walls, it is important to the avoidance of cracking that joint widths are large enough to ensure that walls cannot expand into edge contact at the joint (Figure 4.5 on page 38) although sealants will have failed well before this point is reached.

Since restraint in parapet or other freestanding walls is assumed to be small, the brickwork can be assumed to experience the full theoretical range of size change. Dark coloured brickwork might experience a seasonal temperature range of 85 °C. From $R = \alpha L t$, the unrestrained size change in each 7.5 m length would be about 5 mm. The distribution of this size change must be deduced from the nature of the design and minimum joint clearances calculated accordingly. Some reduction in the value might be assumed on the grounds that the wall will have been constructed at some temperature within the assumed range and above freezing; some risk will be attached to such decisions.

To avoid damage to copings, joints should be carried through them (Figure 4.6 on page 38). Copings, with joints at intervals corresponding to those in the wall, may be of materials with a larger thermal coefficient and lower mass than that of the masonry and so require larger joint clearances. Where size changes in copings are large it is usually preferable for reasons of weathertightness to arrange that their joints are in shear (Figure 4.7 on page 38).

Where masonry is contained by other structures, such as frames, the distribution of size changes is usually more predictable. For example, if the panel of brickwork in Figure 4.3 is dark in colour, it is assumed to experience a seasonal temperature range of 85 °C. From $R = \alpha L t$, the 3 m height will experience an unrestrained size change of 2–3 mm concentrated at the upper joint. This would be, therefore, the least clearance required at that joint to accommodate temperature size change alone, and it would be unwise to assume that restraint might reduce the value. To this value an allowance must be added for possible moisture expansion of the brickwork, for shrinkage and creep in the columns of the reinforced concrete frame, and for the fully compressed thickness of any joint filler incorporated. Also, the vertical joints between the brick infill panels and the columns will be subjected to shear, increasing from zero at the foot to a maximum at the top; structural tying must take account of the varying differential size change.

Cladding
British Standard BS 8200[10] provides data for many aspects of design including those concerned with temperature size changes. It provides thermal absorption coefficients for a range of cladding materials and these are reproduced in Table 4.1 (on page 39). In the absence of specific data, the following approximate coefficients may be used for building materials when dirty:

light surfaces	0.5
medium surface	0.8
dark surfaces	0.9

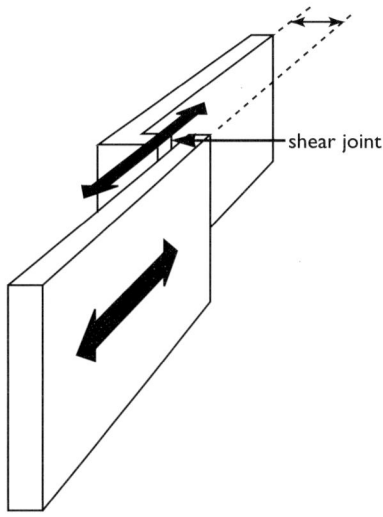

Figure 4.4
A shear joint at an offset in a wall

CRACKING IN BUILDINGS Temperature-induced size changes

Figure 4.5
The purpose of applying sealant to this movement joint has been defeated. The spalled concrete is the result of the expansion of two concrete blocks where the gap between the two has been bridged, or partly bridged, by a fault in casting. Detritus filling the gap can produce the same effect

Figure 4.6
Incorrect detailing in a freestanding wall: the movement joint should continue vertically through the coping

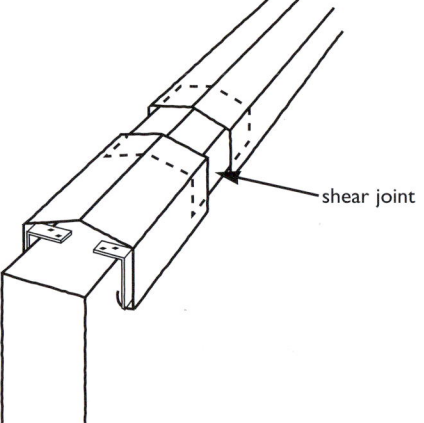

Figure 4.7
A shear joint in a coping

The standard provides equations for calculation of surface temperatures using the relevant absorption coefficients. Two equations are given:

$t_s = t_a + 55a$ for cladding of low thermal capacity
$t_s = t_a + 40a$ for cladding of high thermal capacity

where t_s is the maximum surface temperature in °C
t_a is the maximum air temperature in °C
a is the absorption coefficient of the material.

TABLE 4.1 ABSORPTION COEFFICIENTS OF SOME COMMON CLEAN BUILDING MATERIALS

Material	Absorption coefficient
Black non-metallic surfaces	0.85–0.98
Bricks	0.40–0.70
Stone	0.30–0.55
Concrete	0.65–0.75
Asbestos cement	0.65
Red tiles	0.40–0.65
Dull brass, copper, aluminium	0.40–0.65
Polished brass, copper	0.30–0.50
Highly polished aluminium, chromium	0.10–0.40
Stainless steel, bright polished	0.45
Stainless steel, dull polished	0.65

Cladding with a superficial weight greater than 100 kg/m² is taken as being of high thermal capacity.

The standard suggests that change of size should be calculated in two parts: firstly, for temperatures above the temperature at installation and, secondly, for temperatures below that at installation. In the absence of more exact data, the installation temperature could be taken to be 15 °C. An example of the calculation is shown in the feature panel at the top of the next column.

This assumption for temperature must be treated with caution. As noted earlier, dry construction could proceed even at temperatures below freezing and the assumption of a 15 °C installation temperature would then provide an appreciable underestimate of the expansive part of subsequent size changes.

Site practices
Masonry walls
Provision made in design for temperature size changes can be negated by poor site practices, some of which may be attributable to insufficiently clear specification. Probably the commonest example is that of incorporating mortar in joints intended to permit size change. This can arise from failure to clean out accidental mortar intrusions (Figure 4.8 on page 40) or from the assumption on site that a joint is intended to be mortar filled and only pointed with the specified sealant (Figure 4.9 on page 40).

Sometimes the requirement for a soft joint is overlooked until construction of that part of the wall is complete. A chase is then cut and filled with sealant giving the misleading appearance

---

AN EXAMPLE OF APPLYING BS 8200 DATA FOR CLADDING

Suppose that a particular stone cladding has an absorption coefficient a of 0.4 and the maximum expected air temperature is 30 °C. Then:

$t_s = t_a + 40\,a$
$= 30 + (40 \times 0.4)$
$= 46\ °C$

Suppose that the lowest temperature of the cladding will be −20 °C (page 7). If the temperature of the cladding at installation is 15 °C, the contraction will be over the range +15 °C to −20 °C = 35 °C and the expansion will be over the range +15 °C to +46 °C (the latter calculated as above), a range of 31 °C. Both expansion and contraction can now be calculated from:

$R = \alpha\, L\, t$

with the appropriate thermal coefficient for the stone cladding material and each of the above ranges in turn for t.

If L is the length between joints, the amount by which the installed joint clearance will open and close will now be known, taking account of the distribution of size changes (eg, depending on whether the length L can be assumed to be fixed centrally or at one or other end).

---

that the specification has been met.

Where intermediate joints are incorporated to accommodate size changes, a timber spacer board is sometimes used in order to form a joint of regular width with the possible intention to replace it later with the specified joint filler. The board is difficult to extract without damage to the 'green' brickwork and it remains in place to be concealed behind the sealant. The board is virtually incompressible and damage ensues.

Cladding
From the point of view of minimising cracking, the most significant aspect of work on site concerns the achievement of joints of adequate clearance. This is true for cladding which is exposed to changing air temperatures and to intermittent solar radiation, and especially true for claddings with low thermal capacities (thin or lightweight, for example) or of dark colour.

In an ideal world the designer would provide

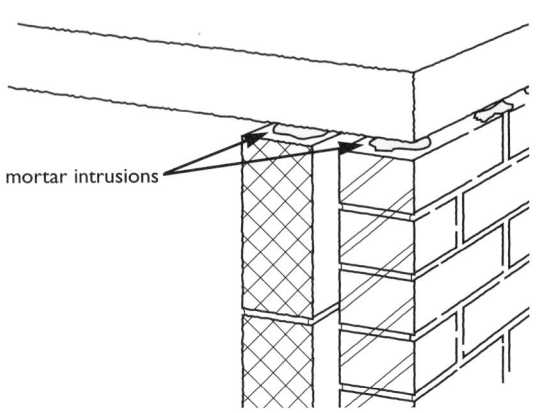

Figure 4.8
Mortar intrusions at the head of brick and block leaves

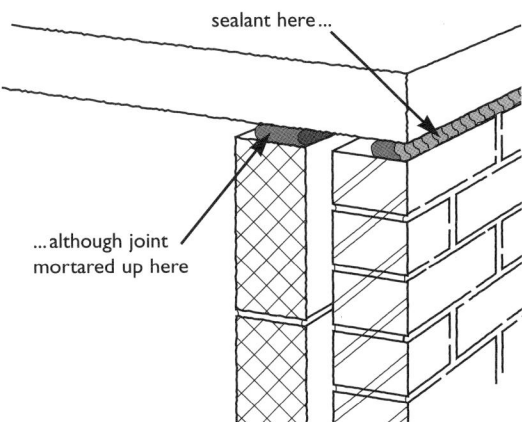

Figure 4.9
An ineffective sealant joint at the head of a brick leaf

site supervisors with joint clearances to be achieved listed against temperatures prevailing at the time of installation. In practice, however, the positioning of cladding components is conditioned by many other factors, not least unavoidable inaccuracies.

While the effects of dimensional inaccuracies can be taken into account in design using the methods in BS 6954:Parts 1–3 [11], the outcome in practice remains heavily dependent on sound site supervision. For example, where it is possible to assemble a run of cladding panels with the aim of approximately equalising all the joints between them, site supervision should ensure that they are not assembled with most joints too narrow and one joint far too wide.

Site supervisors must ensure also that joints intended to accommodate size changes are neither close-butted nor filled, fully or partly, with incompressible material. Similarly, mechanical fixings, intended to permit relative size changes between cladding and supporting structure, must not be constructed so that they provide inadequate reserve either for expansion or for contraction or, worse, so that they will not permit any relative movements to occur.

Mechanical fixings for wall panels are sometimes designed in such a way that both assembly adjustment and provision for size change are incorporated in the same feature – a slot, for example; the dual purpose may not be apparent to site workers, who may assume that the slot or similar feature is provided for assembly adjustment only and overtighten the fixing with no reserve for subsequent movement. Anti-friction devices such as PTFE washers might be omitted if their purpose is not understood. Therefore fixings should be detailed in a way that separates provision for size change from provision for adjustment at assembly, as for example in Figure 4.10.

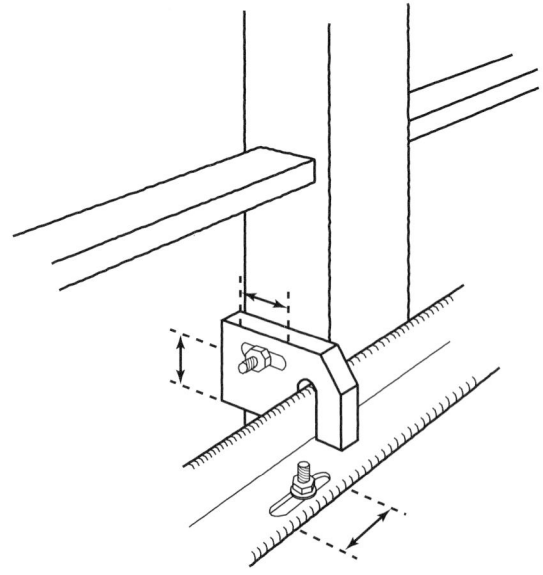

Figure 4.10
Provision for size changes and adjustment (noting that provision for movement is not shown in all three planes)

Temperature-induced size changes CRACKING IN BUILDINGS

Diagnostic principles: walls
Cracks due to thermal contraction are characteristically of uniform width throughout their length and located either at points of greatest restraint or at changes of section. These characteristics are shared by moisture-induced contraction also, and diagnosis should therefore consider first whether the materials concerned are susceptible to moisture-induced size change – either reversible or irreversible, or both. If they are susceptible, it will be necessary to consider the relative amounts of both temperature and moisture-induced size changes.

Tension cracks due to moisture-induced size changes in cement-based products are, generally, more likely to be associated with initial drying shrinkage than with subsequent moisture-induced change. The time of first occurrence of the crack, where it can be established, will be particularly useful diagnostic evidence.

Cracks due to thermal expansion almost always lead to displacement; in such cases identification of the mechanism by which a crack is produced is particularly important (page 13). The extent of displacement will provide evidence as to whether temperature size change, as calculated from the thermal coefficient and an assumption about the possible temperature range experienced, has been responsible. It is often believed that a ratchetting effect can occur due to the accumulation of hard detritus in a crack. If this occurs the displacement might be greater than would be consistent with a theoretical calculation of temperature response alone.

Close inspection of the crack may reveal whether or not substantial amounts of detritus have accumulated. However, a more significant point to remember is that any displacement due to thermal expansion will be related to the largest expansion that has occurred in the building's history of cyclical size changes. Estimation of the amount of size change should, therefore, be based on the worst conditions likely to have been experienced by the building before assessing whether temperature size change has been sufficient to account for the extent of displacement (Figure 4.11).

Typical displacements to be considered are rotation of the ends of intersecting walls, return elevations and short returns within an elevation, oversailing where slip planes occur, misalignment of planes at return elevations, and bowing. (Chapter 2 gives calculation methods.)

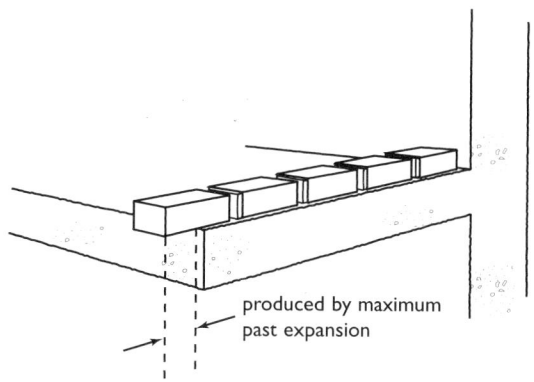

Figure 4.11
Maximum expansion produced by previous worst conditions

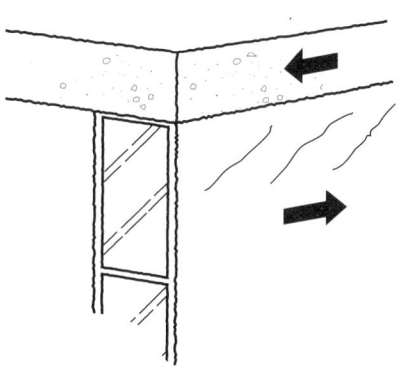

Figure 4.12
Direction of relative size change

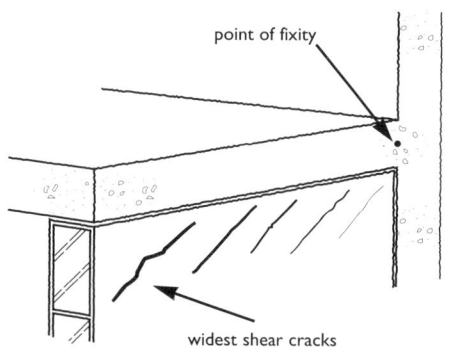

Figure 4.13
Point of fixity

The angle formed by shear cracks following differential size changes between adjacent parts provides a firm indication of the direction in which relative size change is operating (Figure 4.12). Comparison of the width of cracking at various points along the junction firmly indicates their point of fixity from which size changes radiate (Figure 4.13).

Diagnostic principles: cladding
It is in the nature of most cladding supported clear of the main structure that restraints against its thermal contraction are generally small.

CRACKING IN BUILDINGS Temperature-induced size changes

Cracking is therefore more often associated with expansive size changes in the cladding, or with contraction of the main structure if the cladding is contained by it. It further follows that cracking is more likely to be due to compressive forces and is often accompanied by spalling.

Cladding is more responsive to changes in external conditions than the relatively more stable building fabric. It is often of low thermal capacity because of its thin section and it is, of course, more directly exposed to external conditions. The temperature of lightweight claddings can change markedly over a period of a few minutes, with corresponding size changes occurring over a short time span. The following can be relevant to diagnosis of cracking.

- Rapid size changes are unlikely to be accommodated by creep in the materials.
- Shading (eg, by frame members such as mullions and transoms) can produce large local temperature differences with correspondingly high local thermal stress.
- Thin sheet claddings may relieve rapid stress development by bowing ('oil-canning').
- Distortions in the mirrored images of reflective claddings often indicate where local stresses in the claddings occur.

Flat roofs
Design principles

Unless their plan dimensions are very small, flat roofs experience large temperature-induced size changes. The following points are relevant.

- As horizontal elements, the exposure of flat roofs to solar radiation is considerable and, unless shaded (eg, by adjacent taller buildings), they are potentially exposed throughout the day.
- The flat roof coverings in commonest use are dark or very dark in colour and encourage absorption of radiant solar heat by the roof as a whole.
- Except for roofs of more recent design (warm deck and inverted), thermal insulation within the roofs effectively raises the temperature experienced in both the decks and the roof coverings by reducing downward heat dissipation.
- Dark-coloured coverings encourage roofs to radiate heat to the clear night sky in winter so that temperatures of both decks and coverings can fall well below ambient air temperatures.
- Thermal insulation within roofs impedes replacement of the heat lost by radiation to the night sky by heat from the building below, reducing still further the temperatures of both decks and coverings.

All of these effects ensure that flat roofs of conventional cold deck construction experience wide ranges of temperature both daily and, particularly, seasonally.

Where roofs are of heavy construction, perhaps of in situ or precast concrete, their mass slows their dimensional response to temperature change; but while this helps to reduce the diurnal range of temperatures experienced, it has much less effect on seasonal extremes.

Broadly, therefore, flat roofs of cold deck design, whether lightweight or heavy, can be expected to experience the full range of temperature change that can occur in the longer term. There is often comparatively little horizontal restraint given to the roof by the rest of the structure. So if temperature ranges experienced are large, changes of size are also large. Such restraints as exist in the horizontal plane (at the bearing of a roof on intermediate walls, for example) are often insufficient to resist expansive size changes in the roof but offer sufficient restraint to impede subsequent contraction, so inducing cracks in the roof structure. In roof decks that are not monolithic, the same mechanism opens the joints between the units forming the deck.

It is important to remember that materials suitable for use as flat roof coverings are intolerant of localised strain. In principle, a membrane fully bonded over a joint or crack one hundredth of a millimetre wide and which subsequently opens to, say, five hundredths of a millimetre, would need to have an extensibility of 500% to remain unbroken. No such roof membranes exist.

A further relevant point is that cracks and joints in the deck open when thermal contraction is taking place – in other words, the temperature is falling. In these conditions the surface temperatures of flat roofs can fall to well below the air temperatures of winter nights. Decreasing temperature is accompanied by increasing brittleness in roof membranes, and fully bonded membranes fail because their extensibility is least when it is needed most.

Temperature-induced size changes in a roof deck can affect other parts of the structure as well as the covering membrane. These effects can be accentuated if the deck is more effectively restrained at some point on plan than at other locations.

The worst situation arises where a roof of rectangular plan is firmly restrained at or near one end of its major axis, as for example by an abutting building or by a structure such as a lift shaft or motor room penetrating the deck. Temperature size change is then constrained to operate wholly with respect to that point of fixity rather than equally divided about some approximately central point. The cumulative effect is, of course, most marked at places furthest from the point of fixity, and on the longer axis where the displacement will be approximately twice that which would have otherwise occurred.

Thus the distribution of temperature size changes in a flat roof should certainly be taken into account in the design process. Furthermore, there will usually be greater justification (for roofs rather than for other elements) for assuming that restraints will not significantly reduce the predicted unrestrained size change calculated from:

$R = \alpha L t$

In the light of the calculated value for R, and taking account of the effect of restraints – especially of local absolute restraints, as described above – on the distribution of size changes, decisions can be made about the need to subdivide overall size so that size changes do not accumulate to damaging values.

No firm guidance can be given on the limit value for R beyond which it would be necessary to subdivide a major plan dimension. For comparable roof plan sizes and comparable ambient conditions, the calculated value of R for the deck will certainly be larger for cold deck roofs than for warm deck or inverted types, and closer spacing of intermediate joints in the deck is likely to be needed.

As a rule of thumb the introduction of intermediate joints in decks should certainly be considered for cold deck roofs with a major dimension exceeding about 10 m (or 5 m for roofs with absolute restraint at one end of their longer dimension). The locations for any joints required in the deck to accommodate size changes should be selected so as to take account of support locations, the position of any

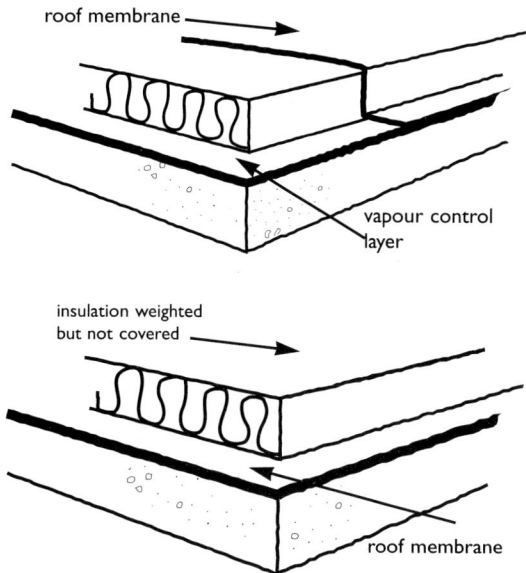

Figure 4.14
A warm deck roof (top) and an inverted roof (bottom)

absolute restraint, the presence of major movement joints in the main structure as a whole, and the locations of any changes of roof shape on plan which could have the effect of a change of section – that is, any departures from simple rectangular plan.

In a warm deck or an inverted roof (Figure 4.14), temperature size changes in the deck will be greatly reduced. In a warm deck roof, however, it is crucial that the vapour control layer is not disrupted by size changes. When these might occur in the deck, so as to open and close its joints, the vapour control layer should not be fully bonded to the deck: otherwise the movements at the joints will split it. Partial bonding should allow movement to take place.

In an inverted roof, the deck and the membrane are protected from solar heat and also prevented from radiating to the night sky; temperature size changes in both will be small or even negligible. The insulation panels, however, if of rigid material, will change size with temperature and allow rainwater to penetrate the open joints so reaching the weatherproof membrane. In some conditions local condensation within the roof may follow. Attempts to seal the 'cracks' between panels are unlikely to succeed and the problem is best dealt with by incorporating additional insulation, in closed cell sheet form, below the main insulation panels.

Practical detailing
British Standard Code of Practice CP 144:Parts 3 and 4[12] recommends that major joints in a flat

roof, including joints required to match movement joints in the main structure of the building, are detailed as capped upstands (Figure 4.15). The recommendations apply to flat roofs with asphalt or bituminous felt membranes, but they could be adopted for other membrane materials also. A similar principle is recommended at perimeter joints where size changes (or deflections) may occur (Figure 4.16).

The sheathing felt laid under asphalt membranes isolates the asphalt from minor size changes in the deck. Built-up bituminous felt membranes can similarly be isolated from minor size changes in the deck by using a perforated base layer which effectively achieves spot bonding rather than full bonding between membrane and deck, so avoiding localised strain.

Membranes fully bonded across deck joints or discontinuities are virtually certain to crack if temperature size changes occur at such points.

Consideration should be given also to possible differing size changes between the deck and any upstand edges or parapets. These differential size changes, acting in shear, can produce ruckling in felt or asphalt upstands (Figure 4.17) if joints are not incorporated in the parapet wall as earlier recommended.

Asphalt membranes on cold deck roofs of larger than small domestic size can be liable to low temperature cracking. Heat loss by radiation to the clear night sky in winter can reduce the temperature of dark surfaces to around −25 °C. The resulting contraction of an asphalt membrane in its relatively brittle state at such temperatures can produce a sudden crack, often accompanied by a loud bang. A solar reflective treatment, besides performing its obvious function, also reduces radiation to the night sky and so reduces the risk of this type of failure. The use of an inverted roof design should avoid the risk of low temperature cracking in an asphalt membrane.

Site practices

Bituminous felt membranes, fully bonded over joints (or discontinuities in the deck) at which temperature size changes accrue, are very likely to crack at those locations for the reasons given earlier. Felt membranes that are turned through a right angle (ie, without employing an angle fillet (Figure 4.18 on page 45) may crack at the angle even if subjected to only very small strain.

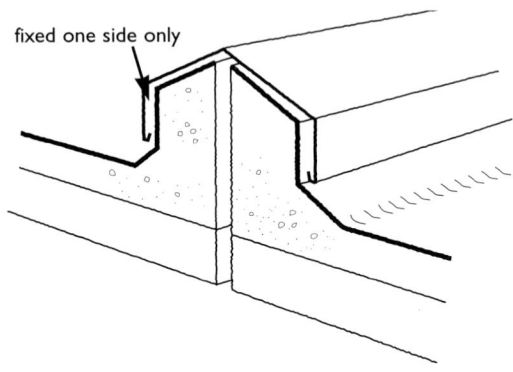

Figure 4.15
A capped movement joint through a flat roof construction

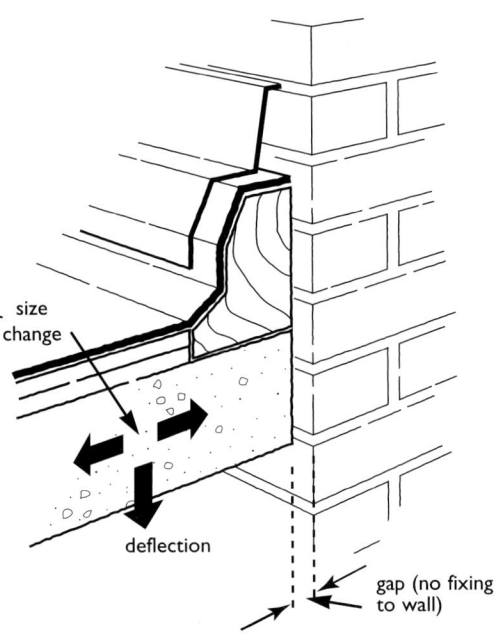

Figure 4.16
A movement joint at an abutment

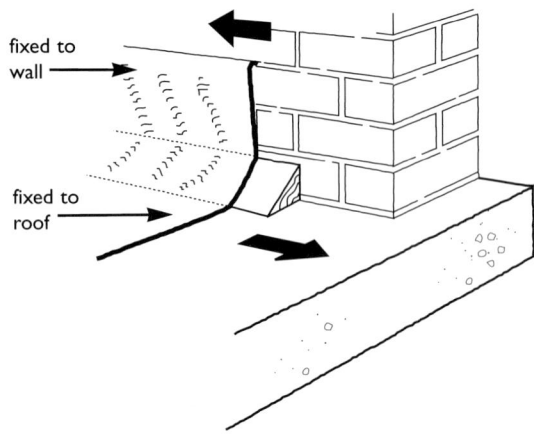

Figure 4.17
Fixing to both roof and wall leads to ruckling at upstands

Temperature-induced size changes CRACKING IN BUILDINGS

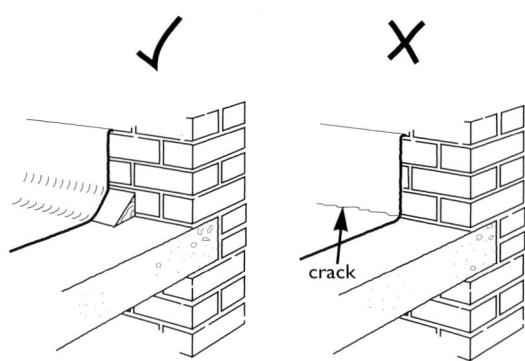

Figure 4.18
Crack prevention by means of a fillet

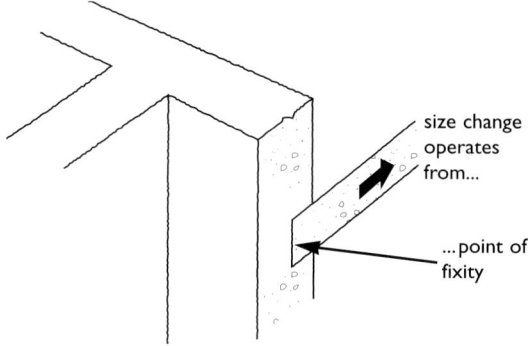

Figure 4.20
The point of fixity for estimation of deck movements

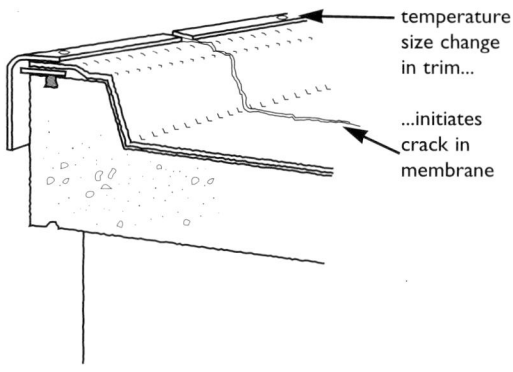

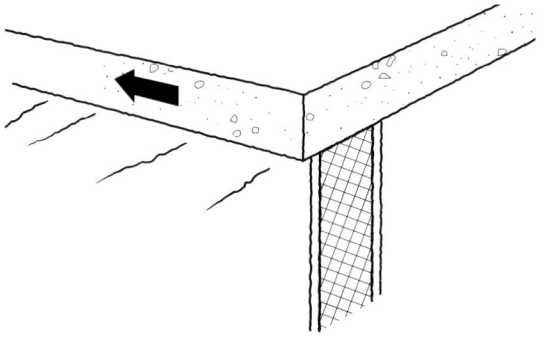

Figure 4.21
Shear cracks at the top of a wall

Figure 4.19
Temperature-induced size changes in eaves trim

Extruded aluminium edge trims, sometimes used with built-up felt roofs, should be securely fixed as close as practicable on both sides of joints in their length; cyclic temperature-induced size changes at the joints in the trim are otherwise very likely to initiate a crack in the membrane at each joint (Figure 4.19).

Perhaps the commonest site practice contributing to cracking in asphalt membranes is overheating the asphalt in the pot. The resulting partial carbonisation of the bitumen content produces a membrane more brittle than it would otherwise have been and thus more likely to crack if subjected to strain.

Diagnostic principles

It is important to relate cracks in flat roof membranes to construction features in the deck and at its perimeter.

Wherever the deck or its structural members (eg, beams and joists) are supported by the main structure, there is likely to be some restraint to temperature size changes in the deck. Where, at such support locations, the deck structure is effectively built into the supporting

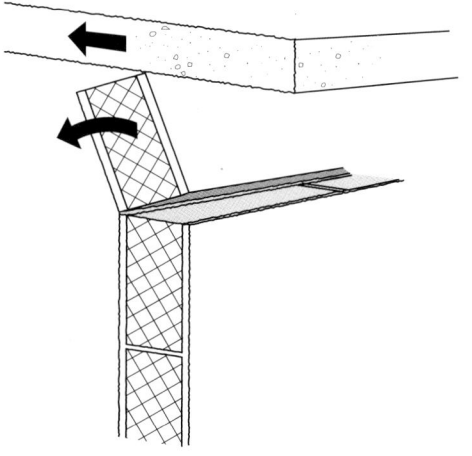

Figure 4.22
Rotation of the top of a wall

main structure, restraint may be considerable or absolute (Figure 4.20). Diagnosis can then make an initial assumption about the way in which temperature size changes in the deck will be predominantly distributed. Note should be taken of any other features (lift shafts, motor rooms, abutting buildings, re-entrants on plan, etc) which might provide a point of fixity and thus increase cumulative size changes elsewhere in the deck.

CRACKING IN BUILDINGS Temperature-induced size changes

The angle taken by ruckles in perimeter upstands will also indicate the direction in which relative size changes are occurring, as also will shear cracks, or rotation, in the tops of intermediate supporting walls in the upper storey (Figures 4.21 and 4.22 on page 45).

A straight-edge applied across cracks in the membrane will reveal whether distortion, or deflection of a deck panel relative to its neighbours, is a more likely cause than linear temperature size changes. On asphalt membranes, temperature size changes can be discounted as the cause of star-shaped cracks or of cracks forming a roughly circular pattern.

Chapter 5

Moisture-induced size changes

Whereas all building materials are affected by temperature change, there are several building materials in common use that are dimensionally unaffected by moisture content changes, or are affected to such a small extent that the practical consequences can safely be discounted. Examples are:

- gypsum plasters (with or without dense or lightweight aggregates or glass reinforcement)
- plaster boards
- metals
- glass
- plastics, although some plastics may shrink irreversibly due to loss of volatile organic constituents

Building materials that experience change of size with change of moisture content can be divided into two groups:

- those that experience reversible size changes, expanding when wetter and contracting when drier
- those that experience irreversible size change

Some important characteristics of the second group should be noted.

- All of those materials that experience irreversible size change experience reversible size change also, the latter being superimposed on irreversible size change so long as it continues.
- Irreversible size changes are largely independent of changes in ambient relative humidity within normal service ranges and are due to progressive chemical or physico-chemical combination with water or water vapour rather than to simple absorption.
- Although irreversible size changes experienced by some materials (such as the irreversible shrinkage of cement-based products or the irreversible expansion of fired clay products) may continue for very long periods, the rate of change diminishes with time and, for practical purposes, can be assumed to be negligible six months after manufacture. However, after a longer period of time (up to 30 years in some cases), irreversible size changes may still contribute to cracking as minor actions which can accumulate to cause major reactions. Thus a diagnosis of the cause of cracking, even though the building is several years old, may still need to consider the amount of irreversible size change that has occurred – albeit now effectively ceased – and its possible contribution to the total of all size changes, both moisture and temperature-induced.

As with thermal responses, moisture-induced size changes in building components can be modified by restraints; restraints can influence the extent and distribution of size changes and may possibly produce distortion. Thus the irreversible shrinkage in a reinforced concrete beam may not significantly reduce its length

Figure 5.1
Poor site practice, such as bricklaying during bad weather, has considerable influence on the incidence of cracking in buildings

CRACKING IN BUILDINGS Moisture-induced size changes

(though increasing its deflection). But the cracks that might have otherwise arisen similarly in an axially-loaded concrete column are closed, in effect, by the axial loading, with corresponding reduction in the height of the column.

The first steps, therefore, are to consider whether reversible size changes will affect the materials and components under consideration; whether these will be superimposed on irreversible size changes; and what might be the effects of restraint – from fixings, friction, reinforcement or loading – on the amount and distribution of size changes or distortion. It should also be kept in mind that moisture-induced size changes happen relatively slowly when compared with temperature size changes. This is the reason why, later in this chapter, it is considered safest to design on the assumption that size changes of these two kinds are additive rather than mutually compensating. It is also advisable, in considering size change at the design stage, to identify the largest likely change in ambient conditions and the corresponding largest likely size change, and then to consider the possible consequences if full provision is not made.

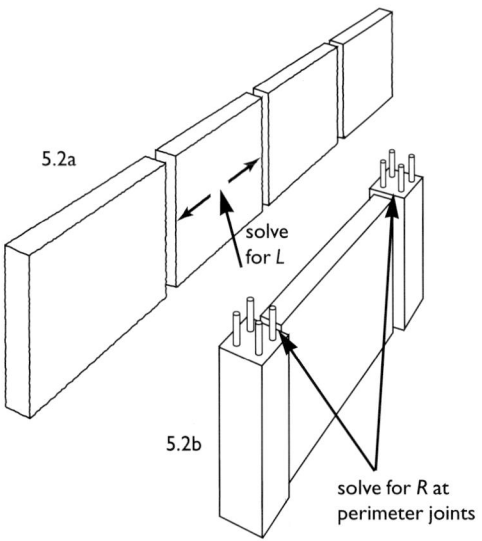

Figure 5.2
Spacing of joints and calculation of perimeter joint width

A point not always made clear in source literature, when distinguishing between reversible and irreversible size changes, is that a part of some reversible size change is effectively irreversible in practice. The usual circumstance is that building materials are wettest at the time of building construction and are never subsequently as wet in service. This is often true of timber components (prompting specific design provision, for example, in brick-clad timber-framed structures) and also true of cement-based materials which at the same time experience truly irreversible size change.

An important practical difference between temperature-induced and moisture-induced size changes should be noted. There is comparatively little that can be done to control the temperature of components at the time of construction on site, but there is much that can be done to control their initial moisture content by protecting them from dampness and rainfall both in storage and during the construction process. Thus site practices potentially have much greater influence on the likelihood of moisture-induced cracking than on temperature-induced cracking (Figure 5.1 on page 47).

Although some shrinkage stress may be relieved by creep rather than cracking if initial drying out of new construction takes place slowly, advice to owners to heat the building only moderately at first is likely to be much less effective in reducing cracking than the adoption by builders of the advice to keep the building as dry as possible during construction.

The considerations dealt with in this chapter relate to provision for moisture-induced size changes only. Design must ultimately take account of other sources of size changes and of building movements.

Walls
Design principles
As with temperature size change, there are two main cases to consider (the same two described on page 34). They apply here to both reversible and irreversible moisture-induced size changes.

In Case 1, the length of the wall under consideration warrants subdivision in order to limit changes of size.

In Case 2, the length does not warrant subdivision, and size changes can be accommodated at the perimeter where joints may in any case be dictated by change of construction at that point.

In both cases the expression:

$$R = \frac{L \times \% \text{ value}}{100}$$

applies where L is the size considered and R the change in that size. In Case 1, the expression is solved for L to determine the appropriate spacing of joints (Figure 5.2a); in Case 2, it is solved for R to determine the change to be accommodated at the joint and so, taking account of other size changes occurring, the design of the joint (Figure 5.2b).

In both cases there are several mechanisms by which cracking might occur:

- compression fracture or distortion in the component considered – as when inadequate provision has been made for its own expansive moisture-induced size change
- tension fracture in the component considered – as when restraint operates on its own moisture-induced contraction (Figure 5.3)
- compression fracture or distortion in the component considered – as when inadequate provision has been made for the contraction of other containing components or structures
- tension fracture in other containing components or structures – as when

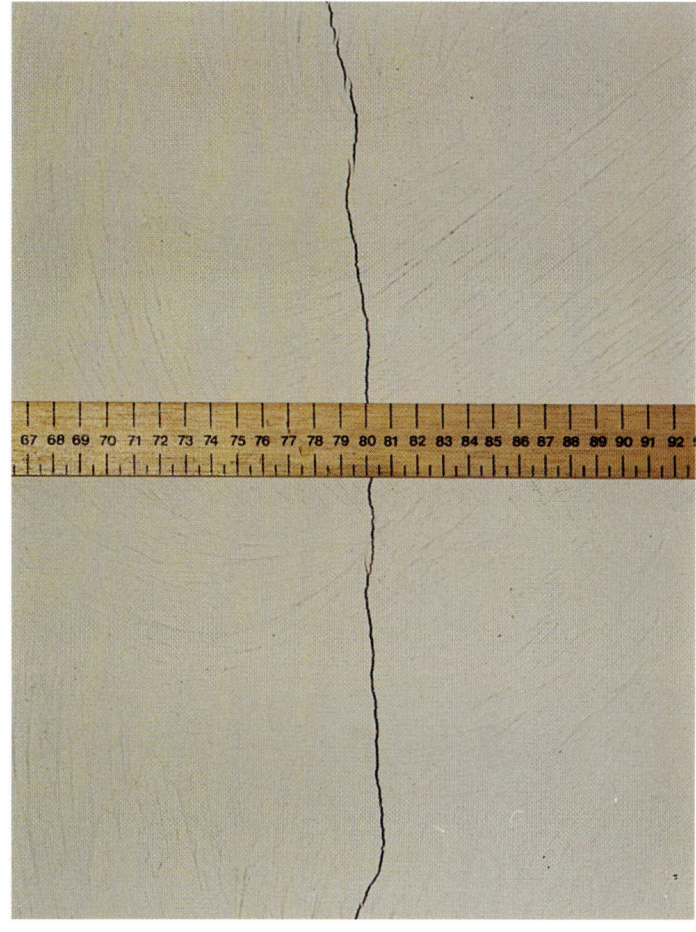

Figure 5.3
A tension fracture caused by moisture-induced contraction

AN EXAMPLE OF CALCULATING MOVEMENT JOINT SPACING IN FIRED CLAY BRICKWORK

If the value for reversible moisture-induced size change given earlier (Table 1.3), 0.02%, is used in the expression:

$$R = \frac{L \times \% \text{ value}}{100}$$

then for a 15 m (15 000 mm) length:

$$R = \frac{15\ 000 \times 0.02}{100}$$
$$= 3\ mm$$

The design of the joint in this example should be such that a total movement range of 3 mm can be accommodated.

As noted earlier in this chapter, the safest design assumption is that this 3 mm range should be added to the 10 mm range calculated on page 36 for temperature-induced size change in the same wall.

For sealant filled joints, the Movement Accommodation Factor expression (page 20) does not require the designer to make an estimate of the distribution of positive and negative size changes within the range of reversible size change.

- inadequate provision has been made for the expansion of the component considered
- shear cracking at planes separating dissimilar materials (Figure 5.4 on page 50)

Other effects, such as displacement of components by size changes in those or other components, also need to be identified: attention is drawn to Chapter 4 and the effects of change of section which apply equally to moisture-induced size changes.

Clay brickwork
The recommendations of BS 5628:Part 3[9] for movement joint spacing in unreinforced masonry are set out in Chapter 4. For fired clay brickwork, a spacing not greater than 15 m is recommended. A calculation is shown in the feature panel alongside.

There is a possibility, in fired clay brickwork, of irreversible moisture-induced expansion. The amount occurring after construction is completed will depend on the age of the bricks at that time. The amount of irreversible

CRACKING IN BUILDINGS Moisture-induced size changes

Figure 5.4
Shear cracking at planes separating dissimilar materials

EXAMPLES OF CALCULATING MOVEMENT JOINT SPACING IN CALCIUM SILICATE BRICKWORK

Taking the higher value and applying it to a 9 m (9000 mm) length of calcium silicate brickwork:

$$R = \frac{L \times \% \text{ value}}{100}$$
$$= \frac{9000 \times 0.05}{100}$$
$$= 4.5 \text{ mm (reversible moisture-induced size change)}$$

The range given for irreversible size change is from 0.01–0.04%. Using the higher value and applying it to the same 9 m length:

$$R = \frac{L \times \% \text{ value}}{100}$$
$$= \frac{9000 \times 0.04}{100}$$
$$= 3.6 \text{ mm (all of which is irreversible contraction)}$$

expansion also varies with the type of clay bricks concerned, the range given earlier in Table 1.3 being 0.02–0.1%. Again using the expression:

$$R = \frac{L \times \% \text{ value}}{100}$$

the corresponding range of possible irreversible expansion in a 15 m length is calculated to be 3–15 mm. Where the bricks concerned are particularly susceptible, and are fresh from the kiln, a 15 mm irreversible expansion might well occur; furthermore, because the expansive force developed is very large, it would be damaging if full allowance were not made for such expansion in design, or if the design allowance were negated by incorrect site actions.

Calcium silicate brickwork
BS 5628:Part 3 recommends movement joint spacing at intervals of 7.5 m and 9 m. Like fired clay bricks, calcium silicate bricks experience both reversible and irreversible moisture-induced size changes. In this case, however, the irreversible change is wholly contraction.

The range given in Table 1.3 for reversible moisture-induced size change is from 0.01–0.05%, depending on brick type. The calculation is illustrated in the feature panel above.

The safest design assumption is that these two size changes are additive; they are also additional to the temperature range of size changes calculated on page 36 as 10 mm for the same 9 m length of wall, giving a total of 18 mm. As with all such safe assumptions, the result may well be over-design for many circumstances arising in practice, but designers must balance this over-design against the risk that smaller provision may result in unacceptable cracks. In masonry walls, however, shrinkage is generally less likely than expansion to produce damage additional to cracking.

Concrete blockwork
BS 5628:Part 3 recommends movement joints at intervals not exceeding 6 m. Taking aerated blockwork as an example, the calculation for size change is illustrated in the feature panel on the opposite page.

Moisture-induced size changes CRACKING IN BUILDINGS

> **AN EXAMPLE OF CALCULATING MOVEMENT JOINT SPACING IN AERATED CONCRETE BLOCKWORK**
>
> Using the upper end of the range given for reversible moisture-induced size change and applying it to a 6 m (6000 mm) aerated blockwork wall:
>
> $$R = \frac{L \times \% \text{ value}}{100}$$
> $$= \frac{6000 \times 0.03}{100}$$
> $$= 1.8 \text{ mm}$$
>
> Although this is the calculated range of reversible size change, the safe assumption is that this will occur as shrinkage from the wet as-constructed state.
>
> The range given for irreversible size change in aerated concrete blockwork is 0.05–0.09%, all of which is shrinkage. Using the same expression and applying the higher value, 0.09%, to a 6 m length yields an irreversible shrinkage of 5.4 mm. Thus the total shrinkage can be assumed to be 7.2 mm in the 6 m length. Again the temperature size change, approximately 3 mm, calculated earlier for the same length, must also be considered. If the assumption is made that all size changes are additive, then the total size change could be assumed to be in the order of 10–11 mm.

Figure 5.5
Shear cracking at the junction of a wall with a ceiling

Practical detailing
Masonry walls
As noted earlier, shrinkage in walls of calcium silicate brickwork or concrete blockwork does not generally produce consequential damage other than cracking. Moreover, the irreversible shrinkage in such walls is of the same order as the entire range of reversible shrinkage. In effect, therefore, the initial irreversible contraction makes, in principle, more than sufficient space for the expansive part of subsequent reversible moisture-induced size changes.

Damage is therefore likely to be confined to tension cracks at junctions at the ends of the wall and possible shear cracking at, for example, the junction with ceilings (Figure 5.5). However, even when joints are provided at the recommended spacings, restraints may produce cracks within the subdivided lengths of walling. Cracking at weak points in walls, for example where windows and doors occur, is very common in shrinkable materials.

Figure 5.6
Cracking at a change of section in a wall

Attention is drawn to the various examples of change of section (given on page 34) which will tend to promote cracks preferentially at such locations (Figure 5.6). Designers can specify sealant filled joints at these positions, but, where

CRACKING IN BUILDINGS Moisture-induced size changes

no change of section occurs, the location of shrinkage cracks generated by restraints will be difficult to predict: tension cracks must then be accepted and made good, if necessary.

Irreversible moisture-induced size change in fired clay brickwork is potentially much more damaging since the change is expansive and large disruptive forces are involved. The consequences of inadequate provision are most apparent when the expanding brickwork is contained by a structure, particularly if the containing structure is itself contracting.

In Figure 4.3, the brickwork panel was calculated to experience a seasonal temperature-induced change of 2–3 mm in its height. The irreversible moisture-induced expansion in its 3 m height, assuming the largest value given for % irreversible expansion is 0.10%, would be 3 mm. The irreversible contraction in the 3 m height of the containing structure, taking the value as 0.08%, could be 2.4 mm. The total reduction in the horizontal joint at the top of the brickwork panel could thus be:

3 mm + 3 mm + 2.4 mm = 8.4 mm.

Failure to allow sufficient space for such an adverse combination of size change could produce substantial damage. Detailing, following the principles shown in Figure 5.7, is usually recommended. The risk of damage is further increased if the brickwork panels are built so that they oversail the supporting nib formed in the perimeter of the in situ concrete floors which are integral with the storey height

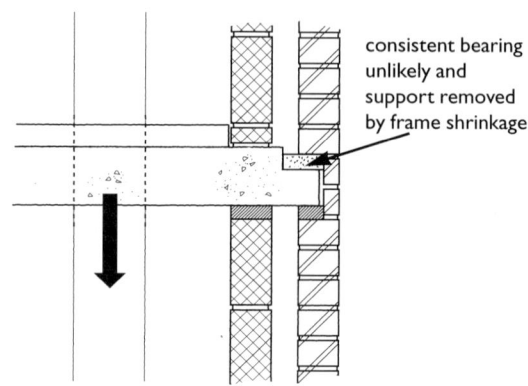

Figure 5.8
Concrete frame shrinkage

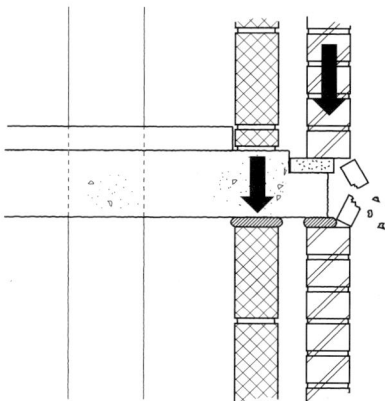

Figure 5.9
Disruption of brick slips

of the frame; for example, to enable brick slips to be used to conceal floor levels (Figure 5.8). Application of the procedures in BS 6954: Parts 1–3[11] for assessing the total effects of inaccuracies will show that the probability that the brick panels will consistently receive adequate bearing is very low indeed. A further problem when brick slips are used in this way arises from the risk that, as vertical shrinkage of the concrete frame removes support from the brick panel above, the load of the panel will be transferred eccentrically to the brick slips (Figure 5.9). The provision of soft joints between the tops of the brick panels and the underside of the floors does not obviate this risk, while the inclusion of such joints also in the brick slip courses might leave them insecure.

Rendered masonry walls
Cement-based renders are potentially subject to both reversible and irreversible moisture-induced size changes. In practice, cracks in renders are rarely if ever attributable to reversible size changes in the render itself. Initial moisture-induced contraction in the render,

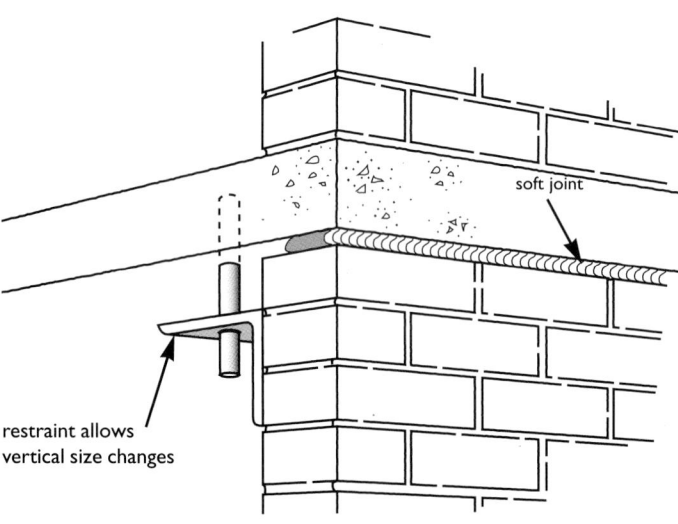

Figure 5.7
A soft joint at the head of a contained brickwork panel

however, is a very common cause of cracks. Renders are also liable to crack as a result of moisture-induced size changes in the substrate to which they are applied. Cracks from this cause are most likely to occur where there is a change of substrate material, the differing materials in the substrate responding differently to moisture changes. At movement joints in the substrate, cracks will be induced in a render that is carried uninterrupted across the joint.

Thus there are two main considerations in design: the choice of an appropriate specification for the render itself, and the adoption of appropriate detailing to avoid cracks due to size changes in the substrate – particularly those that accrue at specific locations.

The principles of appropriate specification include the use of successive coats that are weaker and thinner than preceding coats, while appropriate detailing includes the isolation of the render from the substrate at locations where there is a change of materials in the substrate, and the requirement that any movement joints are to be carried right through to the external face of the render.

Tiled walls
Like other fired clay products, ceramic tiles expand markedly after manufacture and may continue to expand, though at a diminishing rate, for some years thereafter. Crazing of the glaze can indicate such expansion. This moisture-induced irreversible initial expansion, coupled with temperature and moisture-induced changes in the wall to which the tiles are applied, can impose compressive loads in the plane of the tiling, leading to cracks and displacement. Soft joints not less than 6 mm wide are recommended to be incorporated at not more than 4.5 m centres both horizontally and vertically, and at vertical corners in large areas; movement joints in the wall should not be bridged by the tiling.

Timber-framed, masonry-clad walls
Distinction has been made earlier between contained and containing construction insofar as the two cases involve different mechanisms by which otherwise similar size changes produce cracks or other damage. The brick cladding to timber-framed structures is contained, in effect, wherever parts that are integral with the frame intrude into the plane of the brickwork (Figure 5.10). Thus, size changes that reduce the height of the timber frame

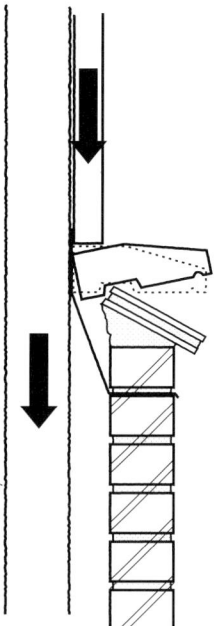

Figure 5.10
Rotation of a sill caused by the shrinkage of a timber frame

potentially impose compressive loads on the brick skin. This can occur, for example, below window sills (where frames and sills are fixed to the structural frame but bear on the brick skin below) and at verges (where the roof structure is carried by the structural frame but bears on the brick skin below).

As indicated in Table 1.3, shrinkage of timber parallel with the grain is small or negligible but across the grain it is appreciable. At some locations, such as intermediate floors, there may be a considerable amount of timber which is across the grain in relation to the height of the building (Figure 5.11 on page 54). To minimise the risk of cracking or other damage, it is necessary to incorporate soft joints below any part which, being fixed to the shrinking structural frame, may bear down unacceptably on the brick skin. The rule of thumb recommendation is to provide 6 mm soft joints in the ground storey, 12 mm in the next storey, and so on.

For similar reasons it is necessary to detail the fixings between lintels and the structural frame in such a way that shrinkage of the frame will not impose a downward loading on the lintel, which would in turn be transferred to the non-loadbearing brick skin.

Site practices
Masonry walls
Materials subject to either irreversible contraction or reversible size changes are very

CRACKING IN BUILDINGS Moisture-induced size changes

likely to crack unless sizes and restraints are adequately small. If such materials form a containing structure, their contraction potentially applies compressive force to the contained construction, with possible consequential damage to either or both. Site practices can markedly influence the likelihood of cracking in all of these cases. Thus, for example:

- reversible size changes are minimised if susceptible materials are kept dry prior to and during construction
- irreversible size changes in materials which also show reversible size changes (ie, in materials that are wetter when installed than subsequently) are minimised if the materials are kept as dry as possible until the building is commissioned
- irreversible shrinkage in cement-based materials is minimised if correctly specified water:cement ratios are not exceeded

Site practices have no effect on the amount of irreversible moisture expansion in fired clay bricks – for example, immersing the bricks in water immediately prior to laying does not remove the risk of moisture expansion; but site practices can negate design provision made for such expansion: where, for instance, soft joints are pointed, or even filled with mortar.

Soft joints may also, incorrectly, be omitted where, in timber-framed construction, shrinkage in the frame might impose compressive loads on the brick skin. This problem often arises where a tiled sill is incorporated below the wood sill of a window frame: the construction sequence makes the joint below the wood sill less accessible for incorporation of the compressible jointing material.

Rendered masonry
As in design (pages 52–53), there are two main aspects to consider: the render itself and the substrate. Site practices impinge mainly on minimising shrinkage in the render and on the practical effectiveness of site work intended to isolate the render from the substrate at locations where size changes accrue.

So far as control of shrinkage of the render is concerned, the key issues are:

- correct mix proportions in successive coats
- lowest water:cement ratio consistent with workability
- clean well-graded sand
- correct curing

Shrinkage cracks are likely if site practices fail to meet any one of these requirements. Such cracks are commonly seen as map-pattern cracking, and, if in renders on clay brickwork, can promote sulphate attack and, hence, further cracking in the render.

Where cracks could be caused by size changes in the substrate, site practices should ensure that movement joints are carried through the thickness of the render with no hard material bridging the joint at any point (Figure 5.12). Otherwise, at changes of substrate material, site work should ensure that the bond between render and substrate is interrupted so that the render is fully isolated from the substrate at those locations (Figure 5.13). Non-corrodible reinforcement should be securely fixed to the substrate on both sides of the isolated area.

Tiled walls
Site practices most likely to lead to cracking or detachment of tiling are:
- failure to incorporate soft joints at adequate intervals
- the incorporation of soft joints of insufficient width

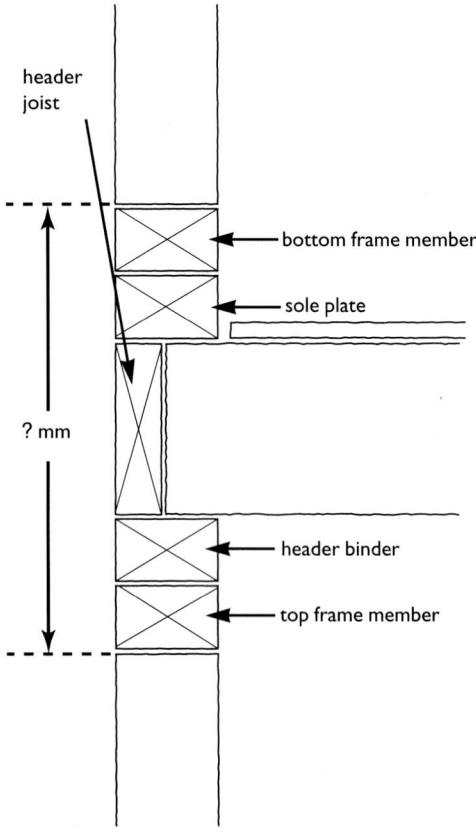

Figure 5.11
Cross-grained timber in the frame of a timber-framed building

Moisture-induced size changes CRACKING IN BUILDINGS

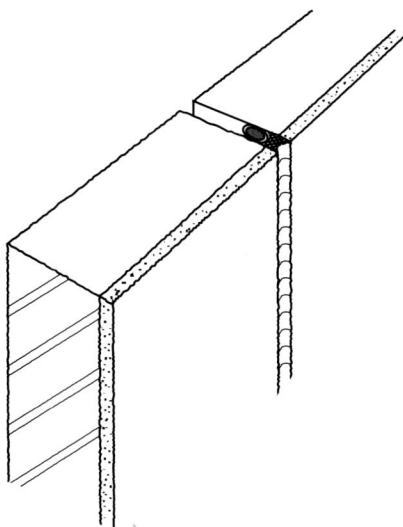

Figure 5.12
A movement joint carried through the render

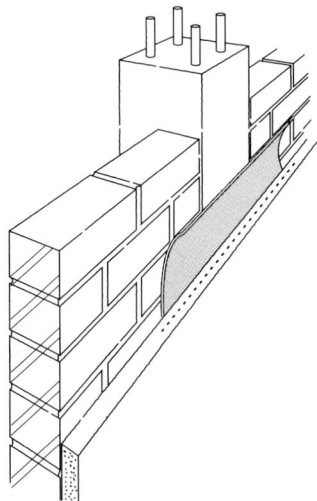

Figure 5.13
Render fully isolated from the column and adjacent brickwork

- the installation of tiling across movement joints in the wall
- failure to allow adequate time for substrate shrinkage before tiling: at least 14 days should elapse before tiling on render, and at least 28 days before tiling on new or newly-plastered walls

Recommendations for other aspects of wall tiling can be found in BS 5385:Part 1[13] and in BRE Defect Action Sheet DAS 137[14].

Diagnostic principles
Moisture-induced size changes
As noted on page 41, cracks due to either temperature or moisture-induced contraction in a wall are characteristically of uniform width throughout their length (Figure 5.3 on page 49). Since all materials are potentially subject to dimensional changes, the first step in the diagnosis of causes is to identify whether the material under consideration is responsive to temperature or moisture changes or both and, if responsive to moisture changes, whether they are reversible or irreversible or both. If the material is potentially subject to irreversible size change early in its life in the building, then evidence as to the age of any cracks will be particularly useful. For example, a crack which first appears within a few months of construction may or may not be due to, say, irreversible initial shrinkage; but a crack first appearing in a building which is 30 years old is most unlikely to have been so caused.

Diagnosis should consider restraints and, in particular, whether the cracked construction forms part of a contained or containing structure: Figures 5.14 and 5.15 (on page 56) demonstrate the containment effect in otherwise exactly comparable circumstances. The effect of the absence of restraints, as for example where a DPC forms a slip plane, should also be considered.

Irreversible expansion is often accompanied by displacement whereas the expansive part of reversible size changes is less likely to produce displacement since earlier contraction will have provided space, to some extent, for subsequent expansion.

Whatever cause or combination of causes is suspected, each must be tested by seeking a mechanism by which it might have produced the observed cracking or damage (pages 13–14). If no reasonable mechanism can be deduced then the supposed cause is unlikely to be the true cause, or at least unlikely to be the major contributor to the observed effects.

Rendered masonry
Map-pattern cracking is characteristic of shrinkage in the render itself. Cracks are often fine enough to be both obscured and sealed by the application of a masonry paint coating, but clearly defined cracks – often the result of applying a strong finish coat over a weak backing coat – may well penetrate the full thickness of the render, with risk of rain penetration and, if on clay brickwork, of sulphate attack (page 63).

Predominantly linear vertical cracks in a render are likely to indicate a change of substrate

CRACKING IN BUILDINGS Moisture-induced size changes

Figure 5.14
Brickwork movement not constrained by an abutment

Figure 5.15
Brickwork movement constrained by an abutment

material at those locations. Since differential size changes in the substrate are likely to continue, such cracks should be raked out and filled with a suitable sealant.

Linear horizontal cracks are unlikely to arise from size changes at a change of substrate materials since vertical loads effectively close cracks that might otherwise have developed in the substrate. They are thus more likely to be due to local compression (eg, of a DPC) in which event the crack is likely to be accompanied by signs of compressive spalling in the render or by relative shear movements at the corresponding plane in the substrate construction; in the latter case there are likely also to be oblique cracks typical of shear. A further possible cause of an isolated horizontal crack is rotation of the wall about that position, as when the upper triangle of a gable end wall is disturbed by longitudinal movements in the roof.

Cracks due to sulphate attack (page 63) and to corrosion of wall ties (page 59) are readily distinguished from those due to moisture-induced size changes.

Floors
Practical detailing for moisture-induced size changes
Chipboard floors
Chipboard experiences large moisture-induced size changes. For example, a change in moisture content from 9–16% produces an expansion in the order of 7 mm in a typical room dimension of 3 m. Provision must be made at perimeters for expansion, with a minimum gap in any circumstances of 10 mm (Figure 5.16). It should also be noted that chipboard can experience a permanent loss of strength if wetted – that is, full strength is not recovered when dry. Requirements for protection of chipboard floors in potentially wet areas are given in BRE Defect Action Sheet DAS 31[15]. Further information is given in BRE Information Paper IP 3/85[16], in BRE Digest 323[17] and in BS 5669:Part 2[18].

Screeded floors
In principle, all cement:sand screeds potentially curl upward at their perimeters or at lines of cracking (Figure 5.17) since drying takes place only from the exposed upper face. In practice, the problem is rare in domestic scale floors, but, in larger areas, it is recommended that consideration should be given to subdivision of screeds into bays if either plan dimension exceeds about 5 m. The risk and extent of curling is influenced by mix

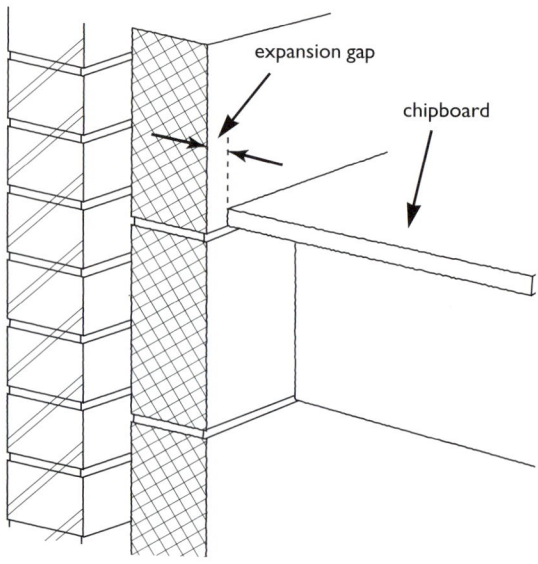

Figure 5.16
A perimeter gap for a chipboard floor

proportions, water:cement ratio, thickness, curing rate and the strength of bonding to the substrate. A common specification would be 1:4 cement:fine aggregate, or, for screeds of more than 50 mm thickness, 1:1.5:3 fine concrete.

Screeds which are subjected to excessive drying shrinkage are likely to develop shrinkage cracks (Figure 5.18).

Screeds which also curl appreciably are likely to crack under load due to lack of support where curled.

Site practices
Chipboard floors
Where moisture-induced size changes are concerned, the crucial requirements of site practice are to ensure that boards are kept dry in site storage and during the construction process, that the correct grade of chipboard (to BS 5669) is used, and that any specified tanking, designed to protect chipboard floors in wet areas, is installed so as to provide complete protection against water.

Screeded floors
Practices that are most likely to increase drying shrinkage and induce curling are:

- the use of sand which is too fine
- having too high a water:cement ratio
- allowing insufficient time for curing

Floor finishes
Wood block and wood panel
Products of this kind, which are intended to provide a superior quality floor finish, are more likely to be well packaged and to receive greater care in storage and during construction, and so be kept dry. Therefore moisture-induced size changes in service, although in principle reversible, are in practice likely to be expansive changes in service. In wood block flooring these occur across the grain and thus in the plane of the floor; expansive size changes can be considerable.

If insufficient provision is made for expansion at perimeters, substantial upward displacement may occur, either as a localised ridge or 'tent' or as uniform bowing over the entire area between restraints (Figure 5.19).

BS 8201[19] recommends normal provision for a perimeter expansion gap of 10–12 mm, but for larger areas the standard recommends that

Figure 5.17
A core has been taken through the intersection of cracking in a cementitious screed. The photograph shows a degree of lift (curling) of several millimetres. Having cracked, the remaining tensile forces in the top of the screed will continue to lift the edges

Figure 5.18
Cracking in a granolithic screed

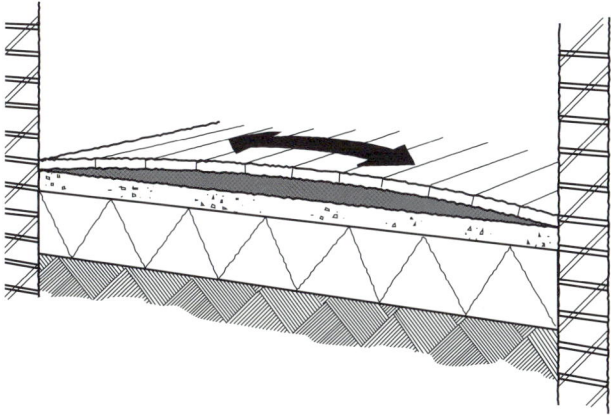

Figure 5.19
Upwards bow on a floor finish

sufficient space be provided between successive individual boards or panels.

Ceramic tiling
BS 5385:Part 3[20] provides guidance on the need for, and the spacing of, joints in clay floor tiling, but encourages designers to calculate the needs specific to the design. The data needed for this are provided in Tables 1.1–1.3 of this book. The standard also provides a number of recommended joint design details.

As with other floorings, particular attention should be paid in design to restraints arising from containment, for example by perimeter walls, steps, plant machinery bases and columns.

Chapter 6
Chemically-induced size changes

Chemically-induced size changes differ from those arising from other causes in two respects.

- The design aim is to avoid their occurrence rather than to accommodate them.
- Their consequences are disruptive cracking rather than cracks of a kind that might be tolerable.

Therefore the general design strategy must be:

- to avoid the use of susceptible materials or of susceptible combinations of materials
- to ensure that service conditions are not favourable to reactions
- to protect susceptible materials, where appropriate, from conditions favouring reaction

The disruptive character of cracking from some kinds of chemically-induced size changes potentially impairs the life and the structural serviceability of affected construction. Designers, specifiers and surveyors should therefore be particularly vigilant in identifying the risks.

Chemical reactions likely to produce disruptive cracking in buildings are:

- corrosion of metals
- sulphate attack
- hydration
- carbonation
- alkali aggregate reaction (including alkali silica reaction)
- conversion in high alumina cement concrete
- degradation of unsound aggregates (usually containing sulphides)

This chapter deals with such chemical reactions and the practical contexts in which they arise, noting some key points relevant to design, diagnosis and repair.

Corrosion of metals
Iron and mild steel are of most relevance in the context of cracking in buildings. The products of corrosion occupy some four to six times the volume of the parent metal. The expansive forces are large and can therefore be extremely disruptive. The consequences are described here in the context of the building components that are potentially susceptible.

Wall ties
Most reported wall tie failures have involved:

- inferior protective coating or none
- inadequate thickness of galvanising
- aggressive (especially black ash) mortars
- permeable (especially lime) mortars
- exposure to severe (eg, marine) environments

However, the problem of wall tie corrosion could eventually affect virtually all cavity-walled structures built before 1981. Furthermore, wall tie corrosion in those structures will not in future be confined to cases of poorly-made ties, nor to aggressive mortars nor conditions of extreme exposure.

BS 5628:Part 3[9] requires the use of corrosion-resistant materials for fixings (including wall ties) above three storeys. The specification of an appropriate austenitic stainless steel for wall ties effectively removes any risk of their corrosion. Site practices, other than the substitution of inferior wall ties for those specified, will then have little if any bearing on the risk of tie corrosion.

Diagnosis
Except for post-1981 construction, almost all cavity-walled buildings built in this century are potentially subject to wall tie corrosion (BRE Information Paper IP 13/90[21]). Cavity walls most at risk are those:

- built in black ash mortars
- exposed to severe weather, especially in marine or industrial environments
- built between 1900 and 1940
- built with vertical twist ties during the shortages which followed the 1939–45 war, or during building booms, especially in the early 1970s
- built with galvanised vertical twist ties where

CRACKING IN BUILDINGS Chemically-induced size changes

the protective coating has exceeded its predicted life. This depends on its thickness, but in most cases will be of 35 years
- built with galvanised wire ties where the protective coating has exceeded its predicted life. This depends on its thickness, but in most cases will be of 20 years
- built with galvanised ties supporting the outer leaf of brick-clad timber-framed construction where the protective coating has exceeded its predicted life. This depends on its thickness, but in most cases will be of 15 years

If the first two conditions listed above are combined with any of the remainder, surveyors should be particularly vigilant.

The surveyor's first need, of course, is to be certain that the wall concerned is of cavity construction. Next, the outer leaf should be examined for regular horizontal cracks at about 300–450 mm spacing; they are usually more evident in the upper parts of the wall and likely to be more clearly delineated on a rendered wall (Figure 6.1). The surveyor should be aware of the possibility that cracks have been repaired or the wall re-rendered.

The cracks are distinguishable from the otherwise very similar cracks caused by sulphate attack on brickwork or rendered brickwork since they occur at vertical wall tie spacing intervals rather than in every (or almost every) bed joint. Corresponding cracks in the inner leaf are rarely found, but some horizontal cracks may be generated where the two leaves are strongly coupled together by structural connections, or where cross walls are bonded to cavity walls. They may also be found in the inner leaves of walls enclosing unheated spaces.

If cracks indicative of wall tie corrosion are found in the external leaf, it is virtually certain that the ties are of the thicker vertical twist variety since wire ties have too little bulk to generate a significant volume of corrosion product unless the mortar joints are very thin. If wall tie corrosion is suspected, it is comparatively simple to locate a tie and to remove a brick, at that location, from the outer leaf, to allow direct inspection of the tie. A decision must be taken on the size of the sample of tie locations to be inspected.

Alternatively an optical probe can be inserted through strategically drilled holes. However, it should be noted that corrosion occurs mainly on the part of the tie that is bedded in the outer leaf, particularly that part within the bed joint and close to the cavity face. If inspection by optical probe does not reveal corrosion of the part of the ties spanning the cavity, this should not be taken as firm evidence without some sampling and direct examination of parts bedded in the outer leaf. Other, generally more expensive, inspection techniques are described in BRE Digest 329 [22].

In all cases of wall tie corrosion, consideration should be given to the possibility that the accompanying expansion has distorted the wall to an unacceptable extent or transferred to the outer leaf loads (eg, roof loads) intended to be carried by the inner leaf. Surveyors should also be alert to the possibility that cavity ties are used where there is no cavity. For example, in some cross-wall housing, separating walls are projected beyond the face of the building and their ends cloaked by a half-brick skin carried up to the full height; stability of the brick skin, in these situations, can depend wholly on the integrity of the ties.

Embedded steel in masonry
Corrosion of embedded steel commonly produces significant, though often localised, disruption of masonry. If alternative, non-corrodible materials cannot be used, the specification of protective treatment is crucial, as is careful detailing aimed to keep the construction dry in the risk area. Protective treatments for structural steel are described in BS 5493 [1].

Figure 6.1
Horizontal cracking of rendered walls

Chemically-induced size changes CRACKING IN BUILDINGS

If corrosion of embedded steel is indicated by localised disruption and rust staining (and drawings, showing specification or the nature of the construction, support the supposition that concealed steel is present), the structural implications of both the corrosion (and weakening) of the embedded steel and the disruption of the masonry will need consideration. In cases where the embedded steel can be safely exposed by partial dismantling, it may be possible to arrest the corrosion process using suitable protective paint systems applied after thorough cleaning of all corrosion products from the steel.

Mild steel reinforcing bars in concrete

In principle, reinforcing steel in concrete is protected against corrosion by the alkali liberated by the process of hydration of ordinary Portland cement. Given adequate depth of cover by good quality concrete of low porosity, reinforced concrete is durable. However, the alkali liberated in the hydration process reacts over time with atmospheric carbon dioxide (carbonation). This reaction proceeds first, of course, at the exposed surface of the concrete. If the concrete is sufficiently dense, and provides adequate depth of cover to the steel, carbonation will not reach the alkali around the steel during the designed life of the structure.

But if the concrete is permeable or cracked, carbonation can reach greater depths and corrosion of the steel may then begin. The rate at which carbonation proceeds is dependent also on both the humidity and the acidity of the environment. The presence of sufficient chloride ions in the concrete can stimulate reinforcement corrosion even in highly alkaline conditions. Corrosion-induced cracking, usually accompanied by spalling, further reduces protection to the steel and thus accelerates the process. Cracking often begins where the depth of cover is least (Figure 6.2).

The specification of concrete of good quality and the requirement for adequate depth of cover are thus crucial elements in the avoidance of reinforcement corrosion. Guidance on the specification of concrete is contained in BS 5328:Parts 1–4[23], in BS 8110:Parts 1–3[24], and in BRE Digests 325[25] and 326[26]. Guidance on required depths of cover for various exposure conditions is given in Table 6.1.

Figure 6.2
Cracking of a column where the depth of cover to reinforcement is least

TABLE 6.1 NOMINAL COVER TO CONCRETE REINFORCEMENT

Condition of exposure	Nominal cover (mm) for concrete grade				
	20	25	30	40	50 and over
Mild: eg, completely protected against weather or aggressive conditions except for brief period of exposure to normal weather conditions during construction	25	20	15	15	15
Moderate: eg, sheltered from severe rain and against freezing whilst saturated with water. Buried concrete and concrete continuously under water	–	40	30	25	20
Severe: eg, exposed to driving rain, alternate wetting and drying, and to freezing whilst wet. Subjected to heavy condensation or corrosive fumes	–	–	50	40	30	25
Very severe: eg, exposed to sea water or moorland water and with abrasion	–	–	–	60	50
Subjected to salt used for de-icing	–	–	50*	40*	25

*Applicable only if the concrete has entrained air

CRACKING IN BUILDINGS Chemically-induced size changes

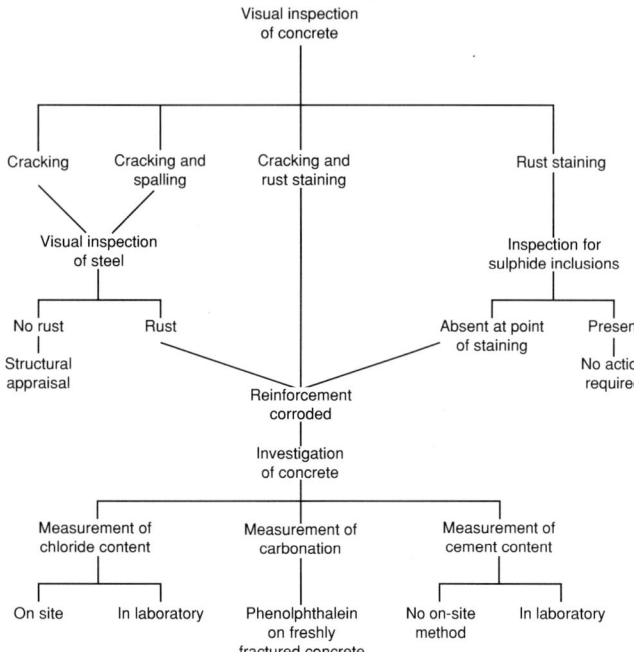

Figure 6.3
A flow chart for the inspection of corroded steel in concrete

Designers should recognise the increased risk arising from the use of slender reinforced concrete components in which adequate cover is more difficult to achieve. Increased risk can also arise in highly reinforced concrete in which congested reinforcing bars may inhibit full compaction of the concrete.

If exposure conditions are severe or the quality of available materials precludes the production of good quality concrete, alternative methods of increasing protection to the reinforcement should be considered. The main techniques available are:

- the addition of corrosion inhibitors to the concrete
- the use of corrosion resistant reinforcement
- the use of reinforcement with applied corrosion protection in the form of metallic or organic coatings
- the use of surface coatings to protect the concrete

Surface coatings are also increasingly specified from the outset to control carbonation and thus preserve the corrosion-inhibiting environment around the steel. Care should be taken to ensure that specified coatings have adequately low permeability to carbon dioxide (BRE Information Paper IP 7/89[27] describes a suitable test procedure).

Site practices which potentially increase the risk of reinforcement corrosion are those which result in:

- reduced cement content
- excessive water in the water:cement ratio
- poor compaction
- contamination of aggregates or of concrete mixes
- incorporation of spacers of incorrect size or material
- insecure location of reinforcement within the formwork

Diagnosis
The diagnosis of corrosion cracking in reinforced concrete should first eliminate causes such as damage by fire, explosion, impact, frost, alkali aggregate reaction, structural overload, settlement, subsidence or other causes. The initiation of corrosion following such damage is unlikely unless the structure has remained unprotected and unrepaired for an appreciable time.

Cracking caused by reinforcement corrosion usually follows the direction of the reinforcement. The location of reinforcement and its depth of cover can be confirmed readily using a cover meter. A structural appraisal may be needed to determine causes if the location of the cracks does not correspond with reinforcement location. Otherwise, if cracking, spalling or rust staining is apparent, investigation can proceed as outlined in Figure 6.3.

It will be obvious from Figure 6.3 that, unless damage has exposed corroded reinforcement to view, once the possibility of reinforcement corrosion is established, the investigation goes beyond mere visual inspection and involves some destructive examination. Such examination will determine the extent of the problem, its possible structural implications and its possible future development. Only then can the need for, and extent of, repairs be considered.

Repairs
Where cracking and spalling are confirmed to be attributable to corrosion of the reinforcement, two important points should first be noted.

- The need for repairs in the future, at locations where cracks are not yet apparent, will very probably arise.

- Each repair, however well executed, carries some risk that it will not be permanent, particularly where both carbonation and high chloride content are found.

Given these limitations, a durable patch repair depends crucially on exposing the full extent of local corrosion, thorough cleaning of the steel to remove all corrosion products (eg, by grit blasting), and the achievement of a close bond between the repair matrix and the existing concrete. Any repair should take account of the structural significance of repair operations as well as the structural soundness achieved by the completed work. There is a wide choice of specialist proprietary repair materials; some generic types and their attributes are described in BRE Digest 265 [28]. Possible alternatives to patch repairs include:

- cathodic protection
- desalination
- re-alkalisation

Sulphate attack

Sulphate salts migrating in solution from neighbouring building materials, or contained in some groundwaters, react with constituents of hardened ordinary (or rapid hardening) Portland cement to produce ettringite or thaumasite. The reaction is accompanied by significant disruptive expansion in mortar, cement renders and concrete (including concrete in aggressive groundwaters). To produce ettringite, the reaction requires the presence of all of the following three constituents:

- soluble sulphate salts
- ordinary or rapid hardening cement
- persistent wetness

To produce thaumasite, the following two further conditions must also coexist with the three constituents:

- a source of carbonate (eg, from the concrete itself)
- low temperatures 5–15 °C

It should be noted that thaumasite can still be formed even though sulphate resisting Portland cement has been used.

Fired clay brickwork

In ideal circumstances, the design strategy must be to avoid the possibility that more than one of the above constituents will be present at any one time. Specification of low-sulphate bricks, or sulphate resisting cement mortar, or the use of detailing which can be relied upon to keep the construction permanently dry, will minimise the risk. Particular attention should be paid to brickwork that will be more than usually exposed to the weather, such as brick parapets and chimneys, where the safest course will be to specify sulphate-resisting cement mortar even though the bricks specified are to be low sulphate (designated 'L' in BS 3921 [29]). Table 13 in BS 5628:Part 3 provides guidance.

Rendered brickwork in exposed locations can be particularly vulnerable to sulphate attack. Renders that have cracked, whatever the reason, allow rainwater to penetrate while also inhibiting subsequent drying. If sulphate attack of the mortar occurs, albeit to a small extent initially, expansion of the bed joints produces horizontal cracks in the render, allowing increasingly copious rain penetration to the brickwork and thus accelerating the process. Brickwork which has been rendered on both faces, such as in freestanding or parapet walls, will become almost permanently wet if the render cracks (as is very likely) since drying from both faces of the brickwork is impeded by the render.

Furthermore, the increase in height of affected brickwork, resulting from the expansion of the bed joint mortar, can disrupt details above it that were intended to exclude rainwater from the brickwork (Figure 6.4 on page 64). In such cases also, since the brickwork becomes not only wetter but also more persistently wet, the conditions favour further sulphate attack and, consequently, disruption of the weathering details.

Diagnosis
Inspection should note the extent to which clay brickwork is exposed to water as direct rainfall, as rainwater run-off from other construction, or as rainwater or groundwater by-passing defective detailing; or water from the ground (which is in contact with, eg, retaining walls), or as leakages from defective rainwater goods or other sources. In dry spells, or on drier brickwork of the same kind elsewhere in the building, efflorescent salts may be visible, indicating the likelihood that the bricks contain sulphates.

Mortar subjected to sulphate attack becomes friable at the surface and, depending on the extent of the attack and the strength of the mortar, shows fairly pronounced cracks along bed joints (Figure 6.5 on page 65). Expansion in the height of the brickwork may be indicated by

CRACKING IN BUILDINGS Chemically-induced size changes

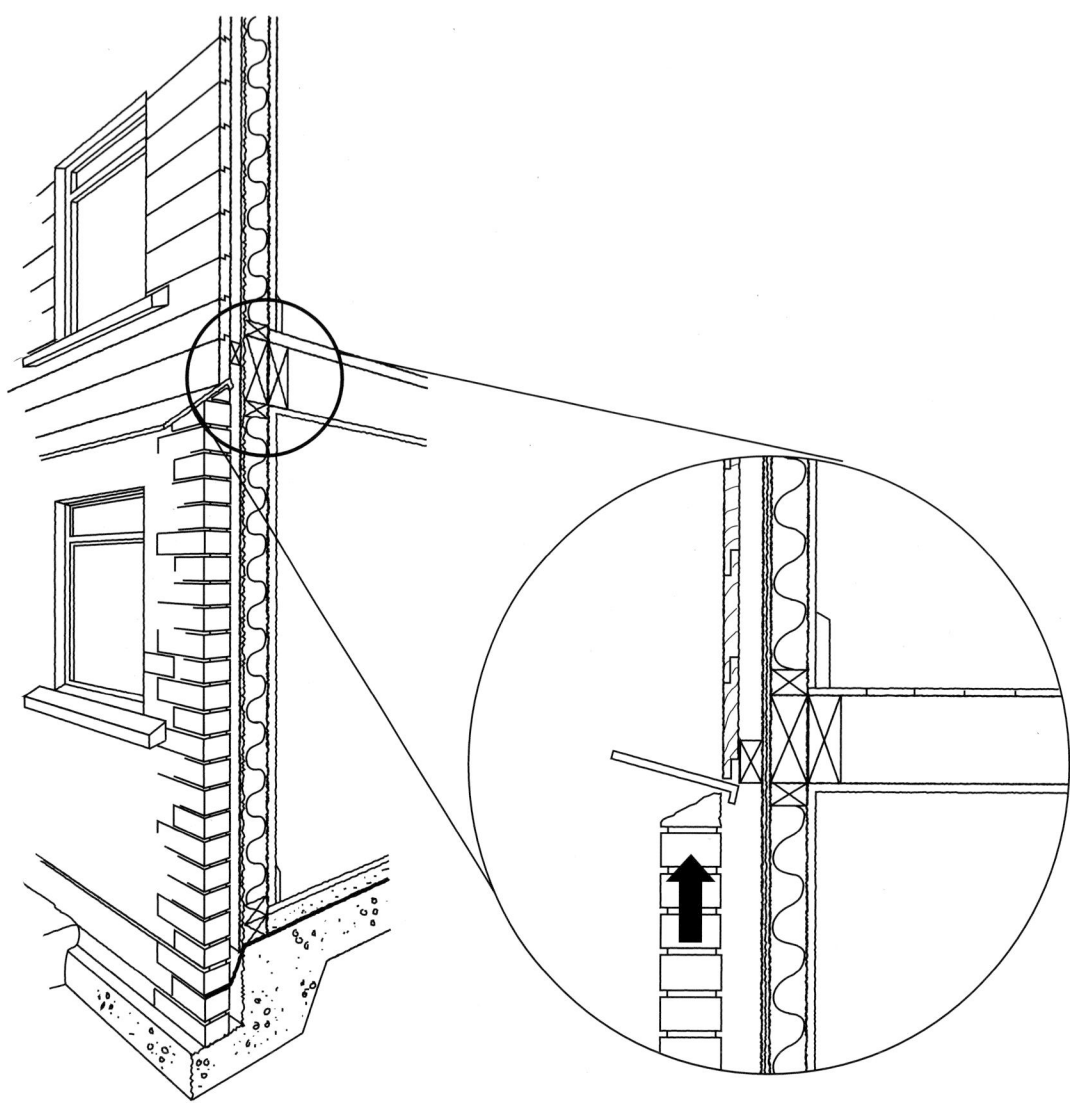

Figure 6.4
Rotation of a tiled sill at first floor level in a timber-framed house

disruption of details at the head of the affected wall or by signs of shear where the sides of the affected wall abut unaffected construction – again particularly near the top of the wall.

Horizontal expansion will be less evident than vertical expansion, partly because there are many fewer perpend joints than bed joints. This might show, for instance, in plan, as oversailing at damp course level. Where affected brickwork is contained, particularly at top and bottom, marked bowing in section may result, though the possibility of other causes (eg, corrosion of strip ties and lack of lateral restraint) should be considered.

The regular horizontal cracking at bed joint intervals that characterises sulphate attack of brickwork will usually be particularly apparent on rendered walls. Such typical horizontal cracking should be distinguished from that caused by corrosion of strip ties (page 60) in cavity walls.

If a confident diagnosis cannot be made by visual examination alone, it will be necessary to have a specialist laboratory examine samples of the affected mortar for the presence of ettringite.

Repairs
Repairs to brickwork damaged by sulphate attack depend on the extent of the damage and its structural significance. If loadbearing brickwork is attacked but is still substantially sound and undistorted, or if a non-loadbearing leaf remains adequately tied, repairs can concentrate on rectifying the causes of persistent dampness. If affected brickwork can thereafter be kept dry, there should be no significant further attack.

Chemically-induced size changes CRACKING IN BUILDINGS

Figure 6.5
Cracks in bed joints in brickwork caused by sulphate attack

Where brickwork has become unstable so that it must be replaced, the repair should incorporate features which will ensure that the replacement brickwork will remain dry. Thus, for example, a replaced brickwork parapet should be provided with a DPC beneath a coping having adequate throated overhangs; the DPC should have support across any cavity, generous overlaps at joints in its length and sufficient width to project slightly on both faces. In some cases it may be possible to apply weatherproof cladding as a means of keeping either the original or the replacement brickwork dry.

Ground-bearing slabs
Sulphate attack on the underside of a concrete ground-bearing slab can occur when fill below the slab contains sulphate salts and the slab is not isolated from the fill by a damp proof membrane (DPM). The reaction produces expansion in the slab in the zone in contact with the fill. The upper zone of the slab is therefore put in tension generating, broadly, a map-pattern of cracking (Figure 6.6 on page 66). Containment of the slab at its perimeter forces the slab to distort into a domed shape which, with time, can become visually very apparent. The corresponding outward thrust at the perimeter often rotates brickwork outward below DPC level. BRE Digest 363[30] provides design guidance where concrete may be subjected to attack by sulphates.

Diagnosis
Sulphate attack can be suspected, particularly if a cracked solid ground floor slab is found in a region where colliery shale fill is likely to have been used; it may also be caused by fill contaminated by sulphates (eg, brick rubble). Again, a floor slab with a pitch mastic or asphalt mastic floor finish is most unlikely to have been provided with a DPM between fill and slab, so that any sulphate-bearing fill would be in direct contact with its underside.

The presence of doming in the slab, detectable with an adequately long straightedge, is characteristic of sulphate attack and so is distinguishable from the effects of heave in clay soil or physical expansion of fill material. If sulphate attack has occurred, a hole broken through the slab will often reveal that it has arched out of contact with the fill. The lower zone of the slab thickness is likely to show white crystalline deposits. If conclusive proof is necessary, samples can be taken for examination in a specialist laboratory for the presence of ettringite or thaumasite.

CRACKING IN BUILDINGS Chemically-induced size changes

Figure 6.6
Cracking in a domestic floor caused by sulphate attack

Repairs
If damage is sufficient to justify repair, the slab must be broken out and removed; it should not be used as hardcore under the replacement slab or in other construction. The fill material, however, can remain (with further compaction if necessary) since the new slab will be laid on a DPM over the fill and turned up at the perimeter. Therefore the new slab will be isolated from sulphates remaining in the fill material. If walls at the perimeter of the slab have been moved outward by its expansion, a judgement must be made about the structural consequences; in most cases the displacement will be small enough to discount.

Hydration and carbonation

Hydration of cement-based products is accompanied by irreversible shrinkage (this latter aspect has been dealt with under the heading 'Moisture-induced size changes' on page 47). However, it is considered again here, together with carbonation, in the particular context of relatively thin sheet products.

As noted earlier in this chapter, under 'Corrosion of metals', the process of hydration of Portland cement liberates free alkali which is converted to carbonate by atmospheric carbon dioxide. Both hydration and carbonation are accompanied by irreversible shrinkage.

In thin sheet manufactured products, such as fibre cement, hydration is normally complete before the products are put to use, but carbonation of the liberated alkali will not be complete. It is generally recommended that holes in sheets for fixings are adequately oversize, to accommodate temperature and reversible moisture-induced size changes in service. The small irreversible shrinkage associated with continuing carbonation is likely also to be accommodated by the oversize provision.

Despite this, cracking can occur in such sheet products if only one face is painted (or otherwise sealed) or if the two faces are painted with films of differing permeability. In these circumstances atmospheric carbon dioxide has readier access to one face. The carbonated face then shrinks in relation to the better protected face, distorting the sheet. If fixings to the building resist that distortion, the sheet may crack, particularly since carbonation is accompanied not only by shrinkage but also by embrittlement.

Embrittlement of asbestos cement sheet with age is a well known phenomenon, and its liability to distort if painted on one face only is perhaps almost equally well known. However, the more recently introduced fibre cement sheet can behave similarly in both respects. Building boards based on calcium silicate may also have sufficient free alkali for differential carbonation (and its attendant distortion and cracking) to occur if carbon dioxide is accessible to one face more than the other.

Diagnosis
If inspection suggests that it is likely that the material concerned is asbestos cement, fibre cement or calcium silicate building board, an investigation should examine the relationship between cracks and fixings and between cracks and the major axis of the board. It will often be apparent whether cracks are associated with restraint by the fixings against linear size change (ie, change in the length of the axis considered) or with restraint against bowing, or with damage caused by overdriving fixings, or with movements occurring in the structure to which the sheets are fixed.

Notice should be taken, particularly, of the presence of a paint or other film on one face of the sheet only. This might well occur, for example, where the underside of an eaves soffit board is painted for decorative purposes while, of course, the inaccessible face within the boxed eaves is not.

Some further inspection may be necessary. For example, amongst a number of sheets which have been fixed at their edges and along the

Chemically-induced size changes CRACKING IN BUILDINGS

Figure 6.7
Typical cracking patterns in unrestrained concrete

centre lines of their major axes, some have cracked and some have not. If, by removing the fixings along the centre line of an uncracked board, it bows up, it will be evident that the board was stressed by those fixings. It would be reasonable to assume, then, that boards which had cracked down the centre line did so because of stressing.

Phenolphthalein indicator solution applied to a newly fractured edge may give a sufficient indication that one face is more carbonated than the other (it will be colourless where the alkali has been fully carbonated and pink where free alkali remains); otherwise laboratory examination may be necessary.

Alkali aggregate reaction

Probably the only significant kind of alkali aggregate reaction in the UK is alkali silica reaction (ASR) and even this is very rare. This reaction takes place between the alkali present in concrete and siliceous minerals contained in some aggregates. It is accompanied by disruptive expansion.

All three of the following constituents must be present for the reaction to occur:

- a sufficiently alkaline solution in the pore structure of the concrete
- an aggregate combination susceptible to attack by that solution
- a sufficient supply of water

Of the few structures in the UK so far affected by ASR, all have been exposed to an external source of water. In the UK, siliceous aggregates are the only type definitely identifiable as susceptible to alkali attack, although the reactivity of such aggregates varies widely.

Apart from the advice to keep concrete structures dry (which would be valid in any event), the interaction between a high-alkali Portland cement and a reactive aggregate can be reduced to acceptable levels by replacement of part of the cement content of the mix by natural pozzolanas, PFA or ground granulated blast furnace slag. However, it still remains current advice to estimate the total alkali content of the concrete. Design advice in this area is complex, and reference to the latest edition of BRE Digest 330 [31] and to The Concrete Society's Technical Report No 30 [4] is advised.

Diagnosis
In unrestrained concrete, the cracking produced by ASR is of the map-pattern type (Figure 6.7) in which areas of fine cracking are bounded by wider cracks. This form of cracking can be distinguished from the similar pattern produced by shrinkage since the latter appears early in the life of the structure whereas cracking due to ASR is not known in structures in the UK less than about five years old.

In reinforced concrete, cracks due to ASR tend to follow the line of prestressing tendons or the line of reinforcing bars; in the latter instance, these cracks can be confused with those produced by corrosion of reinforcement. Certain confirmation of ASR can be provided only by microscopic examination of sections taken from the interior of the concrete.

Repairs
Since the occurrence of damage by ASR is rare, there is little experience of successful repairs. However, since the expansive process involves the continuing take-up of water by the gel formed in the reaction, the only feasible methods of repair will be those aimed at exclusion of water from the concrete.

Clearly the concrete should be as dry as possible prior to the application of any waterproofing treatment; great care must be taken that any changes made with the intention of excluding water are effective and, in particular, do not have the effect either of retaining water within the structure or of inadvertently encouraging increased local wetting.

CRACKING IN BUILDINGS Chemically-induced size changes

Conversion in high alumina cement concrete

Given careful control of cement content and water:cement ratio, high alumina cement concrete (HACC) can have high resistance to attack by sulphate salts and good resistance to acids. However, all HACC undergoes a continuing mineralogical change over time, known as conversion. Conversion is accompanied both by a reduction in strength and an increase in porosity – strength loss is greater in mixes with high water:cement ratios; increased porosity brings greater vulnerability to attack by alkalis. Also, the product of conversion is more vulnerable to sulphate attack than the unconverted material.

The rate of conversion in HACC increases with rise in temperature, both during initial hardening and subsequently in service. Wet service conditions, including high humidities, favour conversion. Moreover, the strength of wet HACC is appreciably lower than that of dry HACC. High rates of conversion produce a disproportionately greater vulnerability to sulphate attack.

Diagnosis and testing
Diagnosis and testing of HACC conversion requires specialist techniques going beyond simple visual examination, and these are described in BRE Digest 392 [32]. However, four general points can be made that may assist surveyors.

- Since conversion rates are greater at high temperatures and humidities, surveyors should be alert, when surveying a building in which such conditions exist, to the possibility that HACC may have been used in its construction.
- The possibility that HACC has been used in unfavourable conditions diminishes in buildings constructed after about the mid-1970s when the potential severity of the problems arising from conversion became widely known.
- The possibility that HACC has been used diminishes in buildings constructed before about 1930 (the first British Standard specification for HACC was published in 1940).
- Virtually all the HACC produced in the UK went into the manufacture of prestressed X or I beams which were cast using a long-line casting process.

Chapter 7

Cracking due to foundation movements

Earlier chapters have dealt in principle with the sources of size changes in building materials and components. Many such sources are likely to be present in any one building and, for practical purposes, temperature and moisture-induced size changes are certain to occur in all buildings. It follows, therefore, that in all normal forms of building construction, they are equally certain to generate cracks. Cracks due to foundation movements (a term which covers subsidence and heave) are relatively uncommon.

However, cracks in a structure, whatever their cause, are often subjectively interpreted as structural cracking in the engineering sense. The attribution of this supposed structural cracking to foundation movements is only a further short step. The over-reaction produced by such a subjective approach is heightened by other considerations, including perceptions of marketability – particularly of a house – and the perceived need, by professionals and by organisations like local authorities, for extreme caution in a litigious climate. Certain judgements in the courts have, no doubt unintentionally, reinforced defensive attitudes: the professional who judges the foundations of a cracked building to be sound may appear to be taking an avoidable risk compared with one who recommends the adoption of remedial measures, whether needed or not.

The consequence of overreaction has been wastage of time and money – the number of occasions when work has been done which need not have been done far outnumber those when work has not been done which should have been done.

An objective approach should first recognise that:

- only a minority of cracks have any structural significance
- only a proportion of cracks of structural significance is attributable to foundation movements
- only a proportion of foundation movements produce appreciable cracking
- only a proportion of cracking due to foundation movements represents a worsening state of affairs

A comprehensive treatment of foundation design is clearly impossible within one chapter of a book such as this - indeed a comprehensive treatment of one aspect is scarcely possible. The aim of this chapter therefore is to alert designers to some relevant published guidance on specific issues, and to try to help surveyors distinguish between cracks that may be related to foundation movements and those that are more likely to have other causes. The reader's attention is particularly drawn to the BRE Report *Foundation movement and remedial underpinning in low-rise buildings*[33] and to *Has your house got cracks?*[34].

Design

The Building Regulations 1991 for England and Wales (as quoted in Approved Document A[35]), require that 'the building shall be constructed so that ground movement caused by swelling, shrinkage or freezing of the subsoil will not impair the stability of any part of the building'.

For normal strip foundations of plain concrete, the thickness must be at least equal to the amount of its projection from the face of the wall or 150 mm, whichever is the greater (Figure 7.1 on page 70). The width of the strip should be that required by Table 12 of Approved Document A.

Where a step occurs in normal strip foundations, the two levels of concrete must overlap by a distance at least equivalent to twice the height of the step or 300 mm, whichever is the greater (Figure 7.2 on page 70). The height of the step must not be greater than the thickness of the concrete strip. There should be no made

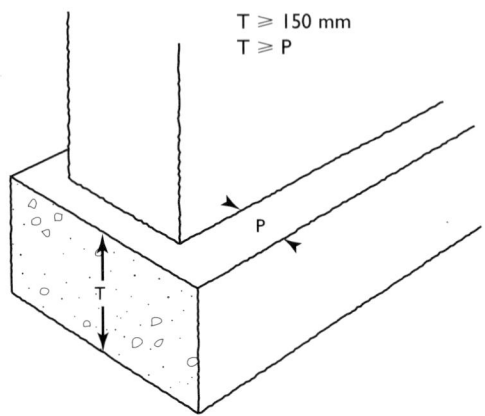

Figure 7.1
Concrete thickness in a normal strip foundation

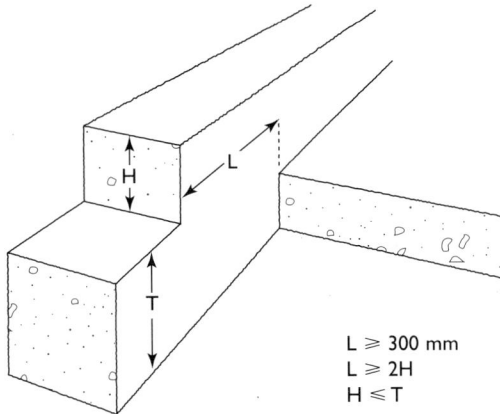

Figure 7.2
Overlap at a step in a concrete strip foundation

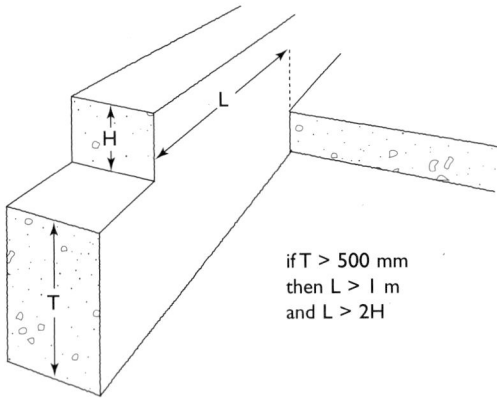

Figure 7.3
Overlap at a step in a deep concrete strip foundation

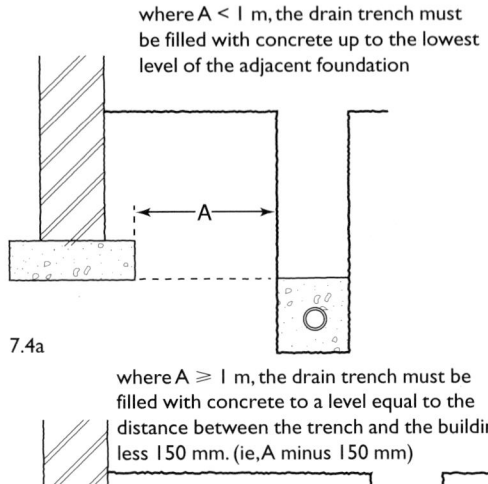

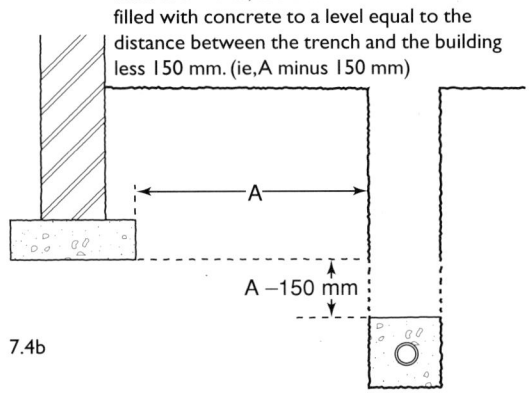

Figure 7.4
Distance of drains from foundations

BS 8103:Part 1[37] recommends that foundations are taken to a depth of 1 m in clays subject to seasonal movements, and to a minimum depth of 450 mm in sands, chalk and other frost-susceptible soils. In the case of frost-susceptible soils, it advises consultation with the approving authority since these soils are widely variable; an increased depth is also advised in upland areas and in areas known to experience long periods of frost.

If a service trench is to be excavated deeper than adjoining building foundations, the Building Regulations (England and Wales) make certain minimum requirements.

For a trench which is less than 1 m from the nearest part of the building, the trench must be filled with concrete up to the level of the base of the foundations (Figure 7.4). Where the trench is 1 m or more from the building, it must be filled with concrete up to a level (below the bottom of the foundations) which is equal to the distance from the building to the trench, less 150 mm.

BS 8004 recommends that short piles be considered as an alternative for low rise dwellings on soft ground, made ground or clays more than 2 m deep, particularly if the water table is high.

ground, and no wide variation of soil type, within the loaded area.

In trench fill or deep strip foundations 500 mm or more thick, BS 8004[36] requires any step to be not greater than the concrete thickness and the lap to be at least 1 m or twice the step height, whichever is the greater (Figure 7.3).

It is well known that later extensions to existing buildings, even if founded at the same depth, are likely to move independently of the existing building and produce cracks at their junction. Less obvious is that extensions that are unheated, including garages and screen wall enclosures, may be subject to frost heave. For both reasons, therefore, it is usually considered safer not to tooth in or otherwise rigidly connect the two buildings but to provide a suitable shear joint. Any crack that subsequently forms at the junction can be flexibly sealed if necessary.

Low rise structures have often been founded on relatively simple foundations at relatively shallow depth where soil tends to be both more variable and more compressible than it is at greater depth. Even for such buildings, therefore, design work should be preceded by site investigations. Competent foundation design for any structure, however small, requires an initial desk study. Valuable information can be obtained at low cost and a desk study is a particularly good investment where the costs of site investigation must be kept low. Table 7.1 (on page 72) provides a desk study check list and Tables 7.2 (on page 72) and 7.3 (on page 73) list some sources of information that can usefully be considered in the desk study.

A desk study should be accompanied by a walk-over survey of the site of proposed construction. The site inspection involves walking across the whole site, making full use of the information obtained during the desk study. It is useful to have available for the site survey, maps and photographs that have been obtained from the desk study, and it is advisable to prepare a special site map at a suitably large scale.

On this map should be marked by hand the geology and other features of special interest that have been noted during the desk study. During the site inspection, the position of any further features (eg, trees, hedges, pits and exploratory holes) can be marked on the map, and the existence of previously identified features can be confirmed. The site inspection should not be confined to the site itself, but should include the surrounding area and its building stock. Any observed defects in buildings close to the proposed site should be investigated and causes established to determine their relevance to the planned building operations.

Information about the following should be recorded: slope angles, ground instability, vegetation, made ground, existing structures and any damage to them, soil and exposed rock, groundwater, evidence of mining, quarrying, and the presence of underlying chalk or limestone. Further guidance is contained in BRE Digest 348[38] which also lists many sources of relevant local information.

While the foundations for most lightly loaded structures can be designed on the basis of experience and using Section 1E of Approved Document A to the Building Regulations for England and Wales, more heavily loaded foundations should be designed in accordance with accepted practice such as is described in BS 8004. Further guidance is also contained in BRE Digest 318[39]. Foundations of any kind on made ground, clay slopes greater than 1 in 10 or other hazardous ground, should be engineer designed. Examples of hazardous ground listed in BS 8103:Part 1 are:

- ground subject to slip or creep which can occur on clay sites with slopes greater than 1 in 10
- areas liable to long term consolidation of the ground, particularly where this may be made-up or reclaimed ground, or where layers of peat are encountered
- areas such as old refuse tips containing material that is subject to spontaneous combustion, chemical change or bacteriological decay, or which include toxic wastes
- ground over or near underground or buried watercourses and ponds
- ground over or adjacent to existing services such as sewers, gas and water mains, and electricity and telecommunications cables
- areas around pits, both natural (such as swallow holes) or due to mining or quarrying; also old bomb craters and soft spots where trees have been removed
- areas liable to subsidence caused by mining or mineral extraction beneath the site
- areas around wells, mine shafts, etc
- old foundations or other concealed constructions
- on clay soils, the close proximity or recent removal of trees or heavy vegetation
- areas liable to flooding or where the water table level is above the expected level of the foundations
- areas where past experience has shown the presence of high sulphate concentrations or other naturally occurring potentially deleterious substances, mainly in clay soils, in sufficient concentration or in circumstances that would cause damage

CRACKING IN BUILDINGS Cracking due to foundation movement

TABLE 7.1 A DESK STUDY CHECKLIST FOR SITE INVESTIGATION FOR LOW-RISE BUILDING

1 Topography, vegetation and drainage
(a) Does the site lie on sloping ground and, if so, what is the maximum slope angle?
(b) Are there springs, ponds, or watercourses on or near the site?
(c) Are, or were, there trees or hedges growing in the area of proposed construction?
(d) Is there evidence of changes in ground level (eg, by placement of fill) or of the demolition of old structures?

2 Ground conditions
(a) What geological strata lie below the site and how thick are they?
(b) What problems are known to be associated with this geological context?
(c) Is the site covered by Alluvium, Glacial Till (Boulder Clay) or any other possibly soft deposits?
(d) Is there available information on the strength and compressibility of the ground?
(e) Is the subsoil a shrinkable clay?
(f) Does experience suggest that groundwater in these soil conditions may attack concrete?
(g) Is there evidence of landslipping either on or adjacent to this site or on similar ground nearby?
(h) Is there, or has there ever been, mining or quarrying activity in this area?
(i) Are there coal seams under the site?

3 The proposed structure
(a) What area will the building occupy?
(b) What foundation loading is expected?
(c) How sensitive is the structure likely to be to differential foundation movements?
(d) What soils information is required for the design of every likely type of foundation?
(e) Is specialist geotechnical skill required?

TABLE 7.2 SOURCES OF INFORMATION AND METHODS OF INVESTIGATION RELATING TO THE TOPOGRAPHY, VEGETATION AND DRAINAGE OF A SITE

Slopes	Current Ordnance Survey maps (at scales of 1:10 000, 1:10 560 and 1:25 000).
Water	Ordnance Survey maps. Water is shown in blue on the 1:25 000 scale maps. This allows easy recognition of springs, ponds, rivers and other drainage features. Water is also coloured blue on some old 6 inch to 1 mile (1:10 560) maps; old maps may show changes in the position of watercourses when compared with more recent maps. On large scale maps - 1:10 560, 1:10 000, 1:25 00 and (in urban areas only) 1:1250 - high groundwater or disrupted drainage may be indicated by symbols for boggy land or heathland.
Vegetation	Aerial photographs provide a permanent record of site vegetation. For a given site, air photograph cover may be available about every ten years from 1946. Aerial photography specialists can estimate the height of trees or bushes from stereo aerial photo cover. Large scale topographical maps will show the positions of hedgelines, woodland and, occasionally, of isolated trees.
Made ground	Comparison of large scale topographical maps made at different dates will allow identification of changes on the site, such as infilling of hollows of old pits, removal of vegetation or demolition of old buildings. Maps of 1:10 560 scale from about 1840 are available; they provide a record of site development over the years. British Geological Survey maps at 1:10 000 scale compiled after about 1965 delimit made ground. (Further information on the BGS is given in Table 7.3.)

TABLE 7.3 SOURCES OF INFORMATION AND METHODS OF INVESTIGATION RELATING TO GROUND CONDITIONS

Geology The geology of a site can be determined from geological maps published by the British Geological Survey, normally at 1:10 560 or 1:10 000 basic mapping scales and summarised at smaller scales of 1:63 360 or 1:50 000. Each 1:50 000 geological map has a descriptive memoir giving details of the local rocks. Where there are extensive Drift deposits, then the Solid geology map is supplemented by a Drift edition. Where the superficial deposits are sparse, a single edition of Solid and Drift will show both forms of geology. Site investigation is generally concerned with superficial deposits. Full collections of maps are available at the London and Keyworth (Notts) offices of the British Geological Survey; Scottish maps are available at the BGS Edinburgh office. Requests for information relating to specific investigations should be made to the BGS Central Enquiry Desk at Keyworth, telephone 01159 363100.

Head (Head is poorly-structured rock debris with shears as well-developed planes of weakness.)
Geological Drift maps, especially at large scale, may identify significant thicknesses (ie, greater than 1–2 m) of Head, but will not necessarily give any warning of the presence of thin layers of such materials; this must be checked by trial pitting. Geotechnical properties of these types of subsoil will be very variable. The only reliable sources of information available at desk study stage are likely to be reports of subsoil investigations carried out in the area (possibly available from the local authority) or technical papers in civil engineering or geological journals.

Landslips Landslipping is outlined on some 1 inch to 1 mile, 1:50 000 and 1:25 000 scale geological maps. It is also shown, where recognised, by notes (such as 'foundered strata') on the manuscript 6 inch County Series maps. Unfortunately, not all landslips have been recorded on these maps. Many can be recognised as hummocky terrain on the ground, but additional help can also be obtained from aerial photographs of the area. The preferred scale for aerial photograph interpretation of this kind is 1:2000, although material up to a scale of 1:10 000 may provide useful information. Black and white photographs are normally the most readily available and should be viewed in stereo for the best effect. To have the best chance of detecting landslipped ground, photographs taken at as many different times as possible should be examined. Landslips are often associated with clays at the base of steep scarp slopes overlain by rocks, such as the Gault Clay at the base of the Chalk in the North Downs. Geological and geotechnical literature are also useful in identifying landslipping.

Chemical attack Chemical attack can result from the natural composition of soils, rock or groundwater in which foundations are to be placed, or from chemical waste in fill (BRE Digest 363*). Much the commonest form is solution by acidic groundwater of carbonate rocks such as Chalk or limestones, or the solution of rocks with carbonate cements. Evaporites, such as rock-salt or gypsum-anhydrite, are dissolved even more rapidly by groundwater. Many British clays contain sulphates. The likelihood of attack from contaminated ground can be determined by looking at topographical maps of different ages to try to detect infilling on the site, and by looking for evidence of old factories or works and trying to assess their potential for pollution (eg, gas works have often left the ground contaminated).

Mining Many different types of mineral have been mined or quarried in the United Kingdom, from flint in Neolithic times to coal at the present. Geological maps at 1:10 000 or 1:50 000 scales will identify layers which contain minerals (eg, coal). Unfortunately there are no abandonment plans of many disused mines. Guidance on the position of old mine shafts can sometimes be obtained from old topographical maps. Other records may be found in local archives such as the County Records Offices, or may be held by central government (eg, the Mining Records Office of the Health and Safety Executive). Other sources include the Catalogue of Abandoned Mines (1928 to 1939) and the Directory of Quarries and Pits (1973). The records of abandoned coal mines are kept at British Coal Archives at Bretby near Burton-on-Trent.

* *Sulphate and acid resistance of concrete in the ground.* BRE Digest 363. Garston, Construction Research Communications Ltd, 1991. ISBN 0 85125 500 0

BS 8004 states that **all made ground should be treated as suspect because of the likelihood of extreme variability.** Where foundations are to be built on new or existing fill, BRE Digests 274[40] and 275[41] provide guidance.

Where, on clay soils, trees are removed prior to construction, precautions can be taken to minimise risk to the building. If trench fill foundations are used, these precautions aim to deal with lateral loads and uplift forces acting on the footings as a result of the swelling of the clay soil that follows tree removal. The steps taken need careful consideration of the relationship between the foundations and the zone of swelling. For example, a compressible layer should be applied only to the side of the trench nearest to the former site of the tree before placing concrete. If applied to the farther side, or to both sides, the resistance to lateral displacement of the foundations will be reduced.

CRACKING IN BUILDINGS Cracking due to foundation movement

TABLE 7.4 RISK OF DAMAGE BY DIFFERENT TREE SPECIES

Note: this Table is a summary of BRE studies of tree damage to buildings. It shows, for each tree species, the tree-to-building distance within which 75% of the cases of damage occurred. The distance is measured between the trunk of the tree and the part of the building nearest the tree

Ranking	Species	Maximum height of tree (H) (m)	Maximum distance for 75% of cases (m)	Minimum recommended separation in very highly and highly shrinkable clays (H – maximum height of tree)
1	Oak	16–23	13	1H
2	Poplar	24	15	1H
3	Lime	16–24	8	0.5H
4	Common ash	23	10	0.5H
5	Plane	25–30	7.5	0.5H
6	Willow	15	11	1H
7	Elm	20–25	12	0.5H
8	Hawthorn	10	7	0.5H
9	Maple, sycamore	17–24	9	0.5H
10	Cherry, plum	8	6	1H
11	Beech	20	9	0.5H
12	Birch	12–14	7	0.5H
13	White beam, rowan	8–12	7	1H
14	Cypress	18–25	3.5	0.5H

A layer of material intended to form a slip plane on the side of the trench fill, so that upward swelling of the clay is less likely to lift the foundations, should not extend below the zone of swelling since this would reduce the resistance to uplift provided by the stable soil below the swelling zone.

Care must also be taken to accommodate swelling soil beneath ground beams and ground-supported floor slabs. BRE Digest 298[42] describes some of the issues.

If it is known that trees will be planted in shrinkable clay soils after construction of the building, consideration should be given to ensuring either a safe separation distance (Table 7.4) or to deepening the foundations to reach a zone that will not be desiccated by the tree roots. If rows of trees are to be planted, the separation distance or the foundation depth should be further increased.

Diagnosis

Since cracks arising from causes other than foundation movements are far more common, the logical first step in any diagnosis would be to seek causes that originate within the building rather than within the ground. It would certainly be time-saving to do so wherever cracks are of uniform width throughout their length.

An estimate of the temperature range experienced and the corresponding size changes in parts of the building bordering a crack can be made using the data in Tables 1.1 and 1.2. Similarly, estimates can be made of the effects of moisture changes, including irreversible expansion (eg, of clay brickwork) and irreversible contraction (eg, of cement-based component parts). Judgements will need to be made about the way in which such size changes will have been distributed before making comparisons with crack widths.

At the same time, evidence should be sought as to whether other factors described earlier in the book (such as sulphate attack, corrosion of wall ties, or live load distribution changes) are or have been operating.

If the evidence indicates that one or more of the temperature sources of size change listed on pages 3–4 is operating, it is unlikely that tension crack widths will exceed 5 mm, and most unlikely that progressive widening will occur whatever their current widths. Cracks of this kind are simply the mechanism by which stress is relieved, and once relieved will not recur. Such cracks can be made good in whatever way is appropriate to the circumstances (repointing, patch plastering, local re-rendering, for example). In a few cases, where dimensions are such that cumulative size changes are fairly large, a flexible crack sealant may be needed.

If the evidence indicates that one or more of the chemical sources of size change listed on pages 10–12 is operating, it is virtually certain that some remedial work will be needed. For example, steps should be taken to rectify defective detailing responsible for persistent wetness where sulphate attack is an ongoing consequence.

When the accumulating evidence suggests that causes of these kinds cannot alone be responsible for the cracks, an investigation should begin to consider whether causes originating in the ground may be contributing. It is suggested that this part of the investigation should begin with a review of any potentially relevant information that is immediately available, recognising that further information may well be needed later.

Next, a preliminary assessment can be made of the likelihood that cracks are related to foundation movements. The first questions, in respect of any one crack in an external masonry wall, should be:

- is it replicated on the inside of the building?
- does it extend below DPC level?
- is the width of the crack tapered? (Figure 7.5)
- is the crack located where maximum structural distortion and structural weakness or change of section coincide?

If the answer to any of the first three questions is no, differential foundation movement is an unlikely cause. The possibility that differential foundation movement is the cause increases with each successive yes answer to all four questions. Further questions then arise.

- Is the crack less than about 5 mm wide and of long standing? If so, differential foundation movements, if they were the cause, were small and may well have ceased; the movements are likely to be attributable to settlement resulting from bedding down under the load imposed by the building.
- Is the crack location related to possible causes of differential foundation movements? For example, to trees on clay soil, to drains that may be leaking (particularly in granular soils), to past or present excavation work, etc.
- Is the building on ground that could give rise to differential settlement? Poor ground, made ground, etc.

It should be noted that settlement – the compression of the ground below the

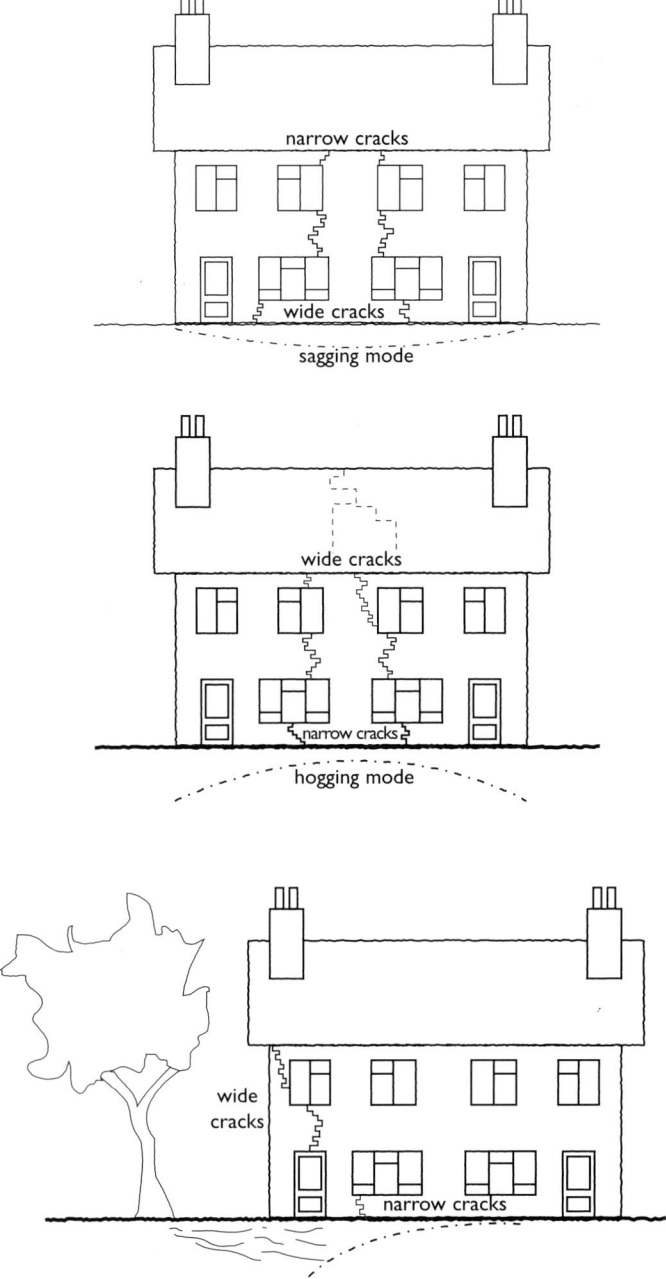

Figure 7.5
Crack patterns associated with different modes of distortion

foundations resulting from the loads applied by the building – is normal and is often accompanied by some differential settlement arising from slight variability in the ground and from slight variations in foundation depths such as may result from removal of soft spots in the trench. Consolidation of granular soils such as sand and gravel under building loads occurs quickly and is often effectively complete at about the time that the construction is completed. Consolidation of clay soils may take some years. Settlement occurs both in foundations and in floor slabs laid on fill that

CRACKING IN BUILDINGS Cracking due to foundation movement

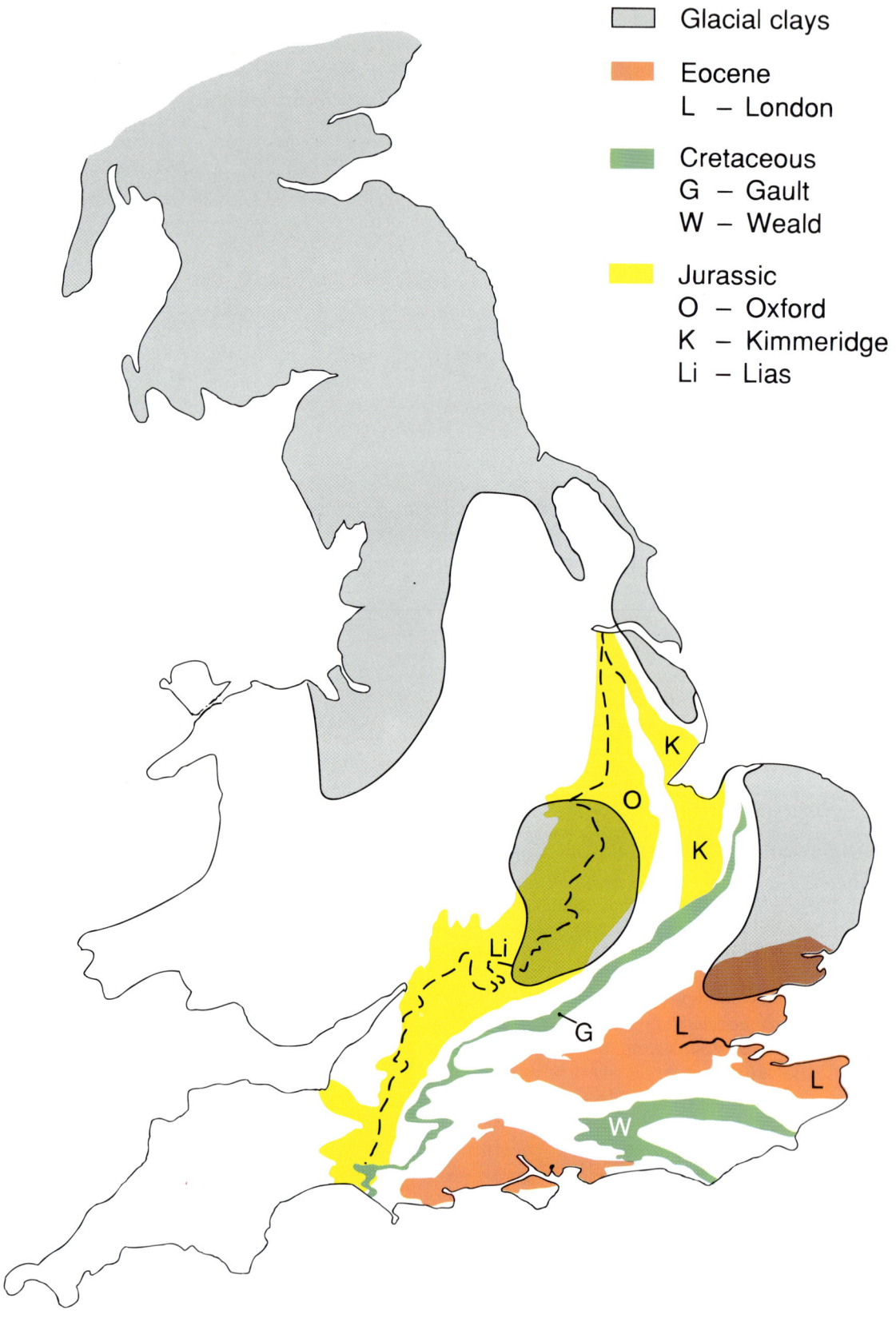

Figure 7.6
Firm shrinkable clays in Britain

was not fully compacted. Cracks in walls and gaps below skirtings can be quite wide and may ultimately need to be made good, but moderate settlement of this kind is commonplace and ultimately self-stabilising.

It is crucially important therefore, if differential foundation movement is suspected, to assess whether movement is progressive. This is particularly true if the soil type is shrinkable clay since some seasonal, reversible movement is certain to occur even under adequately deep foundations.

The general distribution of shrinkable clays in the UK is shown in Figure 7.6, though it should be noted that buildings within the areas indicated as shrinkable clay are not necessarily founded on clay; local soil variations should be checked. Within the areas shown in Figure 7.6, the properties of clay soils vary and foundation depths can be adjusted accordingly depending upon the plasticity index of the clay concerned. British Geological Survey maps and memoirs should be consulted to establish the nature of the ground likely to be encountered at a particular site.

Clay soils, in the top metre or so, undergo marked seasonal volume changes. Buildings founded within this zone move bodily up and down with those volume changes, the amount of movement depending in part on the depth of the foundations. Provided that all parts of the building are founded at the same level and the soil is uniform, no damage necessarily ensues. In practice it is desirable to limit the amount of such bodily movement in a building, in part to avoid damage to drainage systems.

Seasonal wetting and drying and the influence of minor vegetation produce negligible volume change in clay soils at 0.9 m, and this is widely accepted (possibly rounded to 1 m) as the minimum depth for foundations on clay soils. Significant volume changes at depths greater than this are associated with the de-watering effect of the root systems of major vegetation, principally trees. The crucial significance of this is that the volume change is localised and the corresponding movements at foundation level differ at various points under the foundations. It is differential foundation movements that crack buildings.

The removal of trees arrests the de-watering process and the clay returns to its natural water content and thus its natural volume. If the tree is younger than the building then the cracks its presence caused will recover. If the tree is older than the building, or is removed before construction starts, the heave that accompanies reversion of the clay to its original volume can occur comparatively rapidly and is usually more damaging than clay soil shrinkage.

The area of influence of a tree in clay soil varies with the maturity of the tree and its species. Table 7.4 (on page 74) recommends minimum separation distances between various trees and buildings on clay soils.

From the viewpoint of diagnosis, cracks in buildings on shrinkable clay soils are unlikely to be due to differential foundation movements unless:

- parts of the foundations are at different depths
- trees are closer than the distances in Table 7.4
- felled trees older than the building were closer than the distances in Table 7.4
- trees felled prior to construction were closer than the distances in Table 7.4

However, if cracks attributable to differential foundation movement appear to have occurred after a tree has reached maturity, there being no cracks up to that time, it is possible that an exceptionally long dry spell has also had an influence. But cracks will recover when ground moisture contents recover and will not recur to any greater width in future unless an even more exceptionally dry spell occurs. Even in that event, crack recovery ultimately is very probable.

Monitoring movements

Monitoring crack width changes can be useful as a means of confirming any supposed cause. In the context of foundation movements, however, the following questions should be asked.

- Is it certain that cracks are due to differential foundation movement?
- Is it possible that the cause is not self-limiting? For example, that consolidation of the supporting ground is not of a kind that will stabilise before significant damage results.
- Is it uncertain that the movements can be wholly explained by seasonal or reversible changes in the soil, such as will restore cracks?

CRACKING IN BUILDINGS Cracking due to foundation movement

Figure 7.7
Measuring between the shanks of screws with a digital calliper

- Are the cracks not explainable in terms of a cause that can be positively identified by investigation, and dealt with? For example, the growth of major vegetation or washout from leaking drains.

If the answer to any of these questions is yes, monitoring is necessary. Monitoring is always advisable before assuming that movement is progressive. Underpinning or similar work should be considered only when it is established that movement is progressive and will thus produce progressively greater damage.

Crack monitoring requires simple equipment only, but it may need to continue for a year or more before the pattern of movement is apparent. Since the purpose is to monitor changes in crack width, greater measuring accuracy is required than for measuring crack width once only. For the latter purpose it is quite satisfactory to measure crack width to the nearest whole millimetre, perhaps supplemented where necessary with an estimated 0.5 mm or hairline. But measurements for monitoring need to be made to plus or minus 0.1 mm and for this purpose a digital read-out calliper is particularly convenient.

Measurement is made, at intervals, of the change in the distance between screws fixed firmly in the wall. An arrangement of three screws set in a right-angled triangle, as described in BRE Digest 343[43] and shown in Figures 7.7 and 7.8, provides an opportunity to measure rotational as well as linear movements and this is especially desirable where suspected differential movement is monitored. BRE Digest 344[44] describes how these data may be extended by measurements of levels and of verticality to provide a very full picture of building distortion (Figure 7.9). BRE Digest 251[45] provides guidance both on the causes of movements and the assessment of their consequences.

Alternative, but less satisfactory, methods of measurement which are sometimes used, include:

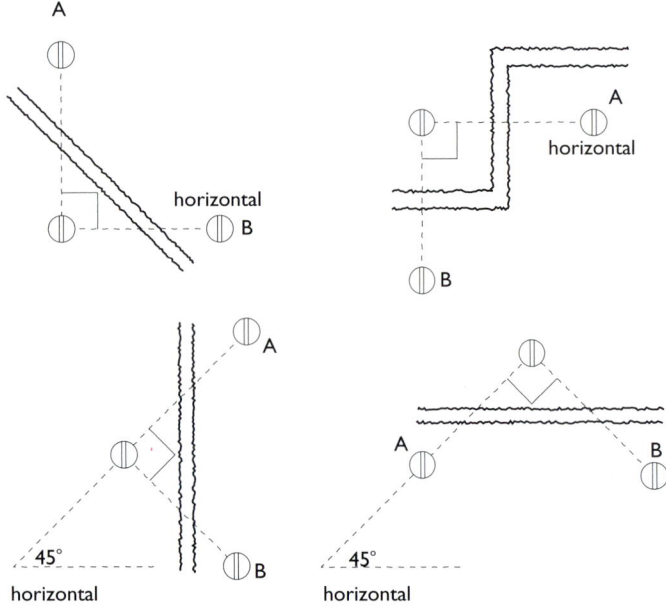

Figure 7.8
Examples of layouts for the 'three screws' monitoring method for various orientations of cracking

Figure 7.9
Measuring the out-of-plumb of a wall. The hand contact with the wall enables the rule to be steadied and prevents it from touching the line

TABLE 7.5 SUMMARY OF SOLUTIONS TO TREE ROOT PROBLEMS

Age of tree	Future growth of tree		Solutions	Advantages	Disadvantages
Older than building	More growth expected		Prune tree	Cheap; tree preserved	Not a permanent solution; effect uncertain
			Deepen foundations (underpinning)	Permanent solution; tree preserved?*	Expensive, disruptive
			Remove tree	Permanent solution to shrinkage problem; cheap	Possible heave problem
	No more growth		Repair damage after winter recovery; perhaps prune	Simple, cheap	May be slight recurrence in a future dry summer if tree not pruned
Younger than building	More growth expected	Very dry summer caused cracking	Repair damage after winter recovery; perhaps prune tree	Simple, cheap	May be slight recurrence in a future dry summer if tree not pruned
		Normal summer caused cracking	'Slight' cracking†; repair cracks after winter recovery	Simple, cheap	Probable future cracking
			Worse cracking: remove tree	Cheap, permanent solution	Loss of amenity
			prune tree	Easy, cheap	Maintenance required
			deepen foundations	Probably permanent solution	Expensive, disruptive; may not give indefinite protection
	No more growth		Repair damage after winter recovery; perhaps prune	Simple, cheap	Possible slight recurrence in very dry summer if tree not pruned

* Most underpinning firms insist on tree removal in the terms of the guarantee
† *Assessment of damage in low-rise buildings with particular reference to progressive foundation movement.* BRE Digest 251. Garston, Construction Research Communications Ltd, 1981 (revised 1993); ISBN 0 85125 461 6

CRACKING IN BUILDINGS Cracking due to foundation movement

- Demec points: dimpled discs fixed on each side of the crack. A limited range of movement only can be measured
- direct measurement across the crack with a steel rule or magnifier and graticule. The magnifier provides greater accuracy than the hand held rule
- glass tell-tales. These give little indication of how much movement is taking place, and are easily vandalised. Their use should be avoided
- plastics tell-tales. These include graduated plates, on which measurements may be made directly. They have limited accuracy

Remedial work

If monitoring shows that foundation movement is indeed progressive (remembering that this circumstance is very rare), remedial measures will be needed to arrest the continuing movement. Remedial measures based on a misunderstanding of what is happening will be ineffective. BS 8004 recommends that 'before underpinning is resorted to, the fullest possible investigation should be carried out by an experienced and competent person to determine whether an underpinning procedure will achieve the object intended'.

Where it seems certain that differential foundation movements are occurring and are attributable to the effects of trees on clay soils, the most reliable confirmation involves digging a trial pit close to the foundations and close to the centre of movement. From this trial pit can be determined:

- the depth of the foundations
- the condition of the clay
- the presence of roots

Table 7.5 (on page 79) sets out a summary of possible remedial actions.

Where repairs to masonry walls are to be undertaken, BRE Digest 359[46] provides guidance.

Appendix 2 offers an approach for diagnosis of the causes of cracking. It attempts to take account of unavoidable practical limitations – particularly of constraints on time – which apply to most valuation surveys. For that reason its use must be augmented by professional judgement, even in that context.

Chapter 8
Other causes of cracking

Earlier chapters have dealt with cracking due to physically and chemically-induced size changes in the component parts of buildings, and due to changes in support from the ground. A few sources of cracking do not fall into these broad categories and are described briefly here.

Mechanical damage

Where the risk of mechanical damage – for example, from vehicles – is real, steps can be taken to protect the structure by means of substantial barriers. Vehicle barriers may be provided from the outset or installed subsequently when the occurrence of damage makes the need apparent. For surveyors, however, the problem of diagnosis may not be straightforward.

Where vehicles, including heavy transport (eg, articulated vehicles with overhanging loads) have frequent access to the curtilage, the risk of mechanical damage to the building may be high, but cracks cannot of course be ascribed to impact on those grounds alone. Fortunately for the surveyor the damage caused by vehicles is usually very recognisable, the evidence consisting mostly of sidelong scuffing, scraping and traces of paint, or of marked punching inward accompanied by broadly vertical and horizontal cracks at the perimeters of the indented area (Figure 8.1).

Surveyors should keep in mind that buildings which are not now especially liable (or liable at all) to vehicle damage may have been so in the past. Former farmhouses, together with barns, stables and other outbuildings, converted for ordinary domestic occupation have often suffered impacts from farm vehicles in the past. At the date of an inspection, however, there may be no route by which vehicles could reach the building and thus no cause to suspect impact as a cause of damage.

Damage by vehicles quite commonly occurs at the salient corners of a building and, on the basis of its corner location, can be mistakenly ascribed to foundation movement. Evidence of abrasion at or near a corner may provide a clue to the true cause (Figure 8.2 on page 82). If obviously longstanding repairs to such damage have remained intact or show no more than the expected shrinkage cracks associated with the repair, this will support the view that the cause has not been foundation movement.

Vibration damage

Vibration damage is mentioned here because it is thought to occur more often than it does. Authenticated cases of damage to buildings by vibration caused, for example, by machinery or by nearby traffic are indeed very rare. It is likely that, even in those rare cases, the vibration has been at most a trigger cause of damage to parts which were already stressed almost to the point of cracking – stressed, that is, by causes other than vibration.

The belief that buildings can be damaged by, for example, traffic vibration rests to a large extent

Figure 8.1
Impact damage to a wall by a vehicle

CRACKING IN BUILDINGS Other causes of cracking

Figure 8.2
Abrasion to brickwork at a quoin

on the remarkable ability of the human frame to detect vibrations of very small amplitude. The resulting subjective impression of heavy vibration may then be reinforced by other factors, such as windows rattling in the air currents created by a vehicle passing in close proximity. Therefore surveyors should not be unduly influenced by occupants' subjective impressions of vibration: the true cause of any cracking ascribed to vibration is virtually certain to lie elsewhere.

Indirect damage

Damage quite often occurs in a part of a building as a result of changes experienced by some other part. Examples mentioned earlier in the book include the rotation of the tops of partitions in upper stories by size changes in a flat roof, and the rotation of the ends of walls by size changes in walls that abut at right angles. In such cases the relationship between cause and effect is fairly close, and thus readily apparent. Sometimes, however, indirect damage is remote from the primary cause and the connection less readily apparent.

For example, distortion of a brick gable following sulphate attack on a chimney contained in it can, by way of the roof purlins, rotate the opposite gable, producing in it a horizontal crack at the base of the gable triangle.

Figure 8.3
Frost attack in clay brickwork

The importance in diagnosis of establishing the mechanism by which a supposed cause has produced damage is accentuated where there is such remote relationship between cause and effect.

Frost damage

Frost damage is most frequently seen in parapets, freestanding walls and retaining walls. The common factor is the exposure of such walls to continual wetting and to temperatures lower than are experienced by most other parts of heated buildings.

Frost damage falls only marginally under the heading of cracking since the consequence is usually spalling of the surface rather than cracking of individual units or of the construction generally. Clay brickwork includes materials most vulnerable to frost attack (though

Figure 8.4
Flaking and spalling caused by frost attack

resistance is high for some classes given in BS 3921[29]). Frost attack of clay brickwork is quite distinctive (Figure 8.3). The characteristic flaking and spalling (Figure 8.4) is unlikely to be confused with the effects of other causes.

Appendix 1

Category of damage	Description of typical damage *Ease of repair in italic type*
0	Hairline cracks of less than about 0.1 mm which are classed as negligible. *No action required.*
1	Fine cracks which can be *treated easily using normal decoration.* Damage generally restricted to internal wall finishes; cracks rarely visible in external brickwork. Typical crack widths up to 1 mm.
2	*Cracks easily filled. Recurrent cracks can be masked by suitable linings.* Cracks not necessarily visible externally; *some external repointing may be required to ensure weather-tightness.* Doors and windows may stick slightly and *require easing and adjusting.* Typical crack widths up to 5 mm.
3	Cracks which *require some opening up and can be patched by a mason. Repointing of external brickwork and possibly a small amount of brickwork to be replaced.* Doors and windows sticking. Service pipes may fracture. Weather tightness often impaired. Typical crack widths are 5 to 15 mm, or several of, say, 3 mm.
4	Extensive damage which requires *breaking-out and replacing sections of walls*, especially over doors and windows. Windows and door frames distorted, floor sloping noticeably*. Walls leaning or bulging noticeably*, some loss of bearing in beams. Service pipes disrupted. Typical crack widths are 15 to 25 mm, but also depends on number of cracks.
5	Structural damage *which requires a major repair job, involving partial or complete rebuilding.* Beams lose bearing, walls lean badly and require shoring. Windows broken with distortion. Danger of instability. Typical crack widths are greater than 25 mm, but depends on number of cracks.

* *Local deviation of slope, from the horizontal or vertical, of more than 1/100 will normally be clearly visible. Overall deviations in excess of 1/150 are undesirable.*

Appendix 2

Crack investigation: a suggested approach

The intention of the list which follows is to suggest a structured approach to crack investigation. It is not a check list leading the user to a specific diagnosis, nor is it intended to be the sole basis for an investigation of a building failure.

1 General appraisal of building, its age, use, general construction form and condition, noting any unusual aspect of its materials, structural character, any changes made (especially recent), significant repair (especially recent), potentially relevant information – for example, recent or ongoing nearby construction, nearby trees (proximity, species and maturity) and soil type.

2 Examine cracks, noting:

- width, direction and taper (including direction of taper)
- signs of compression, tension, rotation and shear
- location, distribution and frequency
- potentially-related other cracks, damage, distortion, out-of-plane and out-of-plumb
- whether they correspond with displacements elsewhere (eg, separation of roof tiles, and distortion of window or door frames)
- whether, if in walls, they continue through ground level DPCs
- whether, if in external walls, they are replicated on indoor faces
- whether, if in partition walls, they are replicated on both faces
- whether, if in floor or roof slabs, they are replicated on both faces (where investigation practicable)
- whether there is similar damage in other comparable parts of the same building
- whether there is similar damage in comparable parts of nearby buildings
- their age

3 Make a first assessment of crack severity according to the above list; then seek a cause or causes, but with time and effort at this stage in proportion to supposed severity.

4 Consider whether the cause or causes are likely to lie:

(i) in the cracked part itself or

(ii) in associated parts which impose forces (tension, compression, shear, rotation and bowing) on the cracked part

If (ii), consider whether the forces arise from within the building itself (eg, dead or live loads, deflection, creep or sway) or from external sources affecting the entire building (eg, wind loads or snow loads) or from changes in its support (eg, settlement of made ground, erosion by leakages, poor compaction of fill, seasonal volume changes under shallow foundations in clay, longer term volume changes, mining subsidence, local excavation, swallow holes or landslip).

5 Make an approximate first assessment of temperature-induced size changes and, if applicable, size changes due to initial expansion or contraction and to reversible moisture-induced size changes. Compare estimated changes with crack widths and spacings and joint widths and spacings; relate to any changes of section or of construction or materials. Consider whether any of these size changes are of the right order of magnitude to be responsible, alone, for the cracks; check that the first assessment is consistent with the age of the crack.

6 Consider possible chemical causes: corrosion of metals, sulphate attack on ordinary Portland cement-based products and materials, alkali silica reaction (reactive aggregates), carbonation (of cement-based products, especially sheets). In all cases check whether the constituents for reaction are present and the conditions are favourable to the reaction.

7 For each potential cause identified by these initial assessments, seek a mechanism linking cause and effect. (If that was the cause, how did it produce this effect?) Accept as possible causes only those for which a possible mechanism can be found; re-examine those causes, seeking further evidence that confirms or denies their existence.

8 When a possible cause is thus identified, seek answers to the following questions.

(i) Are the constituents and conditions confirmed to be, or to have been, present?

(ii) Is the mechanism one that can be confidently accepted?

(iii) Is the supposed cause consistent with the evidence obtained at 2 above?

(iv) Is the potential order of magnitude of the cause consistent with the observed effects?

If the answer to either (i), (ii) or (iii) is no, provisionally discount that cause. If the answers to (i), (ii) and (iii) are yes but the answer to (iv) is no, consider whether other causes are also present and contributing to the observed effects.

9 Avoid the assumption that a cause is correctly diagnosed until all other reasonably possible causes have been examined and discounted; do not overlook the considerable probability that more than one cause is operating. Recognise that the first assumptions may be overturned as the investigation yields further information; also that the first general appraisal of the building may later need to be more specific in the light of that further information.

10 If diagnosis indicates foundation movements as a probable cause and there is reason to believe that movements might be progressive, take account of published guidance (eg, BRE Digests 251[45], 343[43] and 344[44]) to decide whether long term monitoring is necessary.

Distinguish between:

- **settlement**: downward movement caused by compression of the ground by foundation loads. Settlement does not crack buildings – only differential settlement potentially does so; damage due to consolidation of poor or made ground usually becomes apparent within the first ten years (BRE Digest 251)

- **subsidence**: downward movement caused by activity in the ground. However, in the absence of trees, progressive subsidence on shrinkable clay (ie, continuing beyond the duration of a drought) is most uncommon (BRE Digest 251). Where clay soils are involved see also BRE Digests 240[47], 241[48] and 242[49]

- **heave**: upward movement caused by activity in the ground

References

1. **British Standards Institution.** Code of practice for protective coating of iron and steel structures against corrosion. *British Standard* BS 5493:1977. London, BSI, 1977. ISBN 0 580 09565 7.

2. **British Standards Institution.** Specification for bitumen-based hot-applied coating materials for protecting iron and steel, including suitable primers where required. *British Standard* BS 4147:1980. London, BSI, 1980. ISBN 0 580 11402 3.

3. **Building Research Establishment.** Concrete cracking and corrosion of reinforcement. *BRE Digest* 389. Garston, Construction Research Communications Ltd, 1993. ISBN 0 85125 619 8.

4. **The Concrete Society.** Alkali silica reaction: minimising the risk of damage to concrete. *Technical Report* 30. Slough, The Concrete Society, 1987. ISBN 0 946691 20 7.

5. **British Standards Institution.** Code of practice for design of joints and jointing in building construction. *British Standard* BS 6093:1993. London, BSI, 1993. ISBN 0 580 21342 0.

6. **British Standards Institution.** Guide to selection of constructional sealants. *British Standard* BS 6213:1982. London, BSI, 1982. ISBN 0 580 12780 X.

7. **British Standards Institution.** Tolerances for building. Recommendations for selecting target size and predicting fit. *British Standard* BS 6954:Part 3:1988. London, BSI, 1988. ISBN 0 580 16506 X.

8. **British Standards Institution.** Guide to accuracy in building. *British Standard* BS 5606:1990. London, BSI, 1990. ISBN 0 580 18657 1.

9. **British Standards Institution.** Code of practice for use of masonry. Materials and components, design and workmanship. *British Standard* BS 5628:Part 3:1985. London, BSI, 1985. ISBN 0 580 14368 6.

10. **British Standards Institution.** Code of practice for design of non-loadbearing external vertical enclosures of buildings. *British Standard* BS 8200:1985. London, BSI, 1985. ISBN 0 580 14352 X.

11. **British Standards Institution.** Tolerances for building. *British Standard* BS 6954:Parts 1–3:1988. London, BSI, 1988. ISBN 0 580 16504 3, 16505 1, 16506 X.

12. **British Standards Institution.** Roof coverings. Built-up bitumen felt. Metric units. *British Standard* CP 144:Part 3:1970. ISBN 0 580 06425 5. Roof coverings. Mastic asphalt. Metric units. *British Standard* CP 144:Part 4:1970. London, BSI, 1970. ISBN 0 580 06325 9.

13. **British Standards Institution.** Wall and floor tiling. Code of practice for the design and installation of internal ceramic wall tiling and mosaics in normal conditions. *British Standard* BS 5385:Part 1:1990. London, BSI, 1990. ISBN 0 580 18012 3.

14. **Building Research Establishment.** Internal walls: ceramic wall tiles - loss of adhesion. *BRE Defect Action Sheet* 137. Garston, Construction Research Communications Ltd, 1989.

15. **Building Research Establishment.** Suspended timber floors: chipboard flooring - specification. *BRE Defect Action Sheet* 31. Garston, Construction Research Communications Ltd, 1983.

16. **Dinwoodie J M.** Wood chipboard - recommendations for use. *Building Research Establishment Information Paper* IP 3/85. Garston, Construction Research Communications Ltd, 1985.

17. **Building Research Establishment.** Selecting wood-based panel products. *BRE Digest* 323. Garston, Construction Research Communications Ltd, 1992. ISBN 0 85125 265 .

18 **British Standards Institution.** Particleboard. Specification for wood chipboard. *British Standard* BS 5669:Part 2:1989. London, BSI, 1989. ISBN 0 580 17728 9.

19 **British Standards Institution.** Code of practice for flooring of timber, timber products and wood based panel products. *British Standard* BS 8201:1987. London, BSI, 1987. ISBN 0 580 14236 1.

20 **British Standards Institution.** Wall and floor tiling. Code of practice for the design and installation of ceramic floor tiles and mosaics. *British Standard* BS 5385:Part 3:1989. London, BSI, 1989. ISBN 0 580 17752 1.

21 **de Vekey R C.** Corrosion of steel wall ties: recognition and inspection. *Building Research Establishment Information Paper* IP 13/90. Garston, Construction Research Communications Ltd, 1990.

22 **Building Research Establishment.** Installing wall ties in existing construction. *BRE Digest* 329. Garston, Construction Research Communications Ltd, 1988 (revised 1993). ISBN 0 85125 282 6.

23 **British Standards Institution.** Concrete. *British Standard* BS 5328:Parts 1–2:1991, Parts 3–4:1990. London, BSI, 1990. ISBN 0 580 20267 4, 20274 7, 18979 1, 18980 5.

24 **British Standards Institution.** Structural use of concrete. *British Standard* BS 8110:Parts 1–3:1985. London, BSI, 1985. ISBN 0 580 14489 5, 14490 9, 14781 9.

25 **Building Research Establishment.** Concrete. Part 1: materials. *BRE Digest* 325. Garston, Construction Research Communications Ltd, 1987. ISBN 0 85125 268 0.

26 **Building Research Establishment.** Concrete. Part 2: specification, design and quality control. *BRE Digest* 326. Garston, Construction Research Communications Ltd, 1987. ISBN 0 85125 269 9.

27 **Davies H and Rothwell G W.** The effectiveness of surface coatings in reducing carbonation of reinforced concrete. *Building Research Establishment Information Paper* IP 7/89. Garston, Construction Research Communications Ltd, 1989.

28 **Building Research Establishment.** The durability of steel in concrete: Part 3. The repair of reinforced concrete. *BRE Digest* 265. Garston, Construction Research Communications Ltd, 1982. ISBN 0 85125 309 1.

29 **British Standards Institution.** Specification for clay bricks. *British Standard* BS 3921:1985. London, BSI, 1985. ISBN 0 580 14642 1.

30 **Building Research Establishment.** Sulphate and acid resistance of concrete in the ground. *BRE Digest* 363. Garston, Construction Research Communications Ltd, 1991. ISBN 0 85125 500 0.

31 **Building Research Establishment.** Alkali aggregate reactions in concrete. *BRE Digest* 330. Garston, Construction Research Communications Ltd, 1988 (revised 1991). ISBN 0 85125 283 4.

32 **Building Research Establishment.** Assessment of existing high alumina cement concrete construction in the UK. *BRE Digest* 392. Garston, Construction Research Communications Ltd, 1994. ISBN 0 85125 625 2.

33 **Hunt R, Dyer R H and Driscoll R.** Foundation movement and remedial underpinning in low-rise buildings. *Building Research Establishment Report*. Garston, Construction Research Communications Ltd, 1991. ISBN 0 85125 459 4.

34 **Freeman T J, Littlejohn G S and Driscoll R M C.** *Has your house got cracks? A guide to subsidence and heave of buildings on clay.* London, Thomas Telford Services Ltd, 1994. ISBN 0 7277 1996 3.

35 **Department of the Environment and The Welsh Office.** The Building Regulations 1991. Approved Document A: Structure (1992 edition). London, HMSO, 1992. ISBN 0 11 752312 7.

36 **British Standards Institution.** Code of practice for foundations. *British Standard* BS 8004:1986. London, BSI, 1986. ISBN 0 580 15166 2.

37 **British Standards Institution.** Structural design of low-rise buildings. Code of practice for stability, site investigation,

foundations and ground floor slabs for housing. *British Standard* BS 8103:Part 1:1986. London, BSI, 1986.
ISBN 0 580 14932 3.

38 **Building Research Establishment.** Site investigation for low-rise building: the walk-over survey. *BRE Digest* 348. Garston, Construction Research Communications Ltd, 1989. ISBN 0 85125 424 1.

39 **Building Research Establishment.** Site investigation for low-rise building: desk studies. *BRE Digest* 318. Garston, Construction Research Communications Ltd, 1987. ISBN 0 85125 240 0.

40 **Building Research Establishment.** Fill. Part 1: Classification and load carrying characteristics. *BRE Digest* 274. Garston, Construction Research Communications Ltd, 1983 (revised 1992). ISBN 0 85125 314 8.

41 **Building Research Establishment.** Fill. Part 2: Site investigation, ground improvement and foundation design. *BRE Digest* 275. Garston, Construction Research Communications Ltd, 1983 (revised 1992). ISBN 0 85125 315 6.

42 **Building Research Establishment.** The influence of trees on house foundations in clay soils. *BRE Digest* 298. Garston, Construction Research Communications Ltd, 1985. ISBN 0 85125 344 X.

43 **Building Research Establishment.** Simple measuring and monitoring of movement in low-rise buildings. Part 1: cracks. *BRE Digest* 343. Garston, Construction Research Communications Ltd, 1989.
ISBN 0 85125 380 6.

44 **Building Research Establishment.** Simple measuring and monitoring of movement in low-rise buildings. Part 2: settlement, heave and out-of-plumb. *BRE Digest* 344. Garston, Construction Research Communications Ltd, 1989. ISBN 0 85125 381 4.

45 **Building Research Establishment.** Assessment of damage in low-rise buildings with particular reference to progressive foundation movement. *BRE Digest* 251. Garston, Construction Research Communications Ltd, 1981 (revised 1993). ISBN 0 85125 461 6.

46 **Building Research Establishment.** Repairing brick and block masonry. *BRE Digest* 359. Garston, Construction Research Communications Ltd, 1991.
ISBN 0 85125 485 3.

47 **Building Research Establishment.** Low-rise buildings on shrinkable clay soils: Part 1. *BRE Digest* 240. Garston, Construction Research Communications Ltd, 1993.
ISBN 0 85125 331 8.

48 **Building Research Establishment.** Low-rise buildings on shrinkable clay soils: Part 2. *BRE Digest* 241. Garston, Construction Research Communications Ltd, 1980 (revised 1990). ISBN 0 85125 377 6.

49 **Building Research Establishment.** Low-rise buildings on shrinkable clay soils: Part 3. *BRE Digest* 242. Garston, Construction Research Communications Ltd, 1980.
ISBN 0 85125 332 6.

Further reading

BRE Digests

- 63 Soils and foundations: part 1
- 64 Soils and foundations: part 2
- 67 Soils and foundations: part 3
- 157 Calcium silicate (sandlime, flintlime)
- 163 Drying out buildings. ISBN 0 85125 126 9
- 196 External rendered finishes
- 217 Wall cladding defects and their diagnosis
- 227 Estimation of thermal and moisture movements and stresses: part 1
- 228 Estimation of thermal and moisture movements and stresses: part 2
- 229 Estimation of thermal and moisture movements and stresses: part 3
- 263 The durability of steel in concrete: part 1 - mechanism of protection and corrosion
- 264 The durability of steel in concrete: part 2 - diagnosis and assessment of corrosion-cracked concrete
- 281 Safety of large masonry walls
- 313 Mini-piling for low-rise buildings
- 315 Choosing piles for new construction
- 342 Autoclaved aerated concrete
- 352 Underpinning
- 353 Damage to structures from ground-borne vibration
- 361 Why do buildings crack?
- 362 Building mortar
- 381 Site investigation for low-rise building: trial pits
- 383 Site investigation for low-rise building: soil description

BRE Defect Action Sheets

- 2 Reinforced concrete framed flats: repair of disrupted brick cladding
- 18 External masonry walls: vertical joints for thermal and moisture movements
- 37 External walls: rendering - resisting rain penetration
- 51 Floors: cement-based screeds - specification
- 52 Floors: cement-based screeds - mixing and laying
- 70 External masonry walls: eroding mortars - repoint or rebuild?
- 71 External masonry walls: repointing - specification
- 72 External masonry walls: repointing
- 75 External walls: brick cladding to timber frame - the need to design for differential movement
- 76 External walls: brick cladding to timber frame - how to allow for movement
- 96 Foundations on shrinkable clay: avoiding damage due to trees
- 122 Windows and doors: reconstituted stone non-structural components; 'plastic' repair using Portland cement mortar - specification
- 128 Brickwork: prevention of sulphate attack
- 129 Freestanding masonry boundary walls: stability and movement

BRE Good Building Guides

- 13 Surveying brick or blockwork freestanding walls
- 23 Assessing external rendering for replacement or repair
- 24 Repairing external rendering

Index

AAR (*see* Alkali aggregate reaction)
Abrasion of walls, 82
Absolute restraints, 43
Absorption coefficients, 39
Abutments, 44, 56
Accuracy, 21, 22, 25
   characteristic, 22, 25
   intrinsic, 25
Acid atmospheric pollutants, 11
Acids, resistance to, 68
Additives, concrete, 11
Adhesion, 13
Adhesive restraint, 28
Aggregates, siliceous, 67
Air pressure differentials, 32
Air temperatures, 40, 42
   ambient, 42
Alkali aggregate reaction (AAR), 12, 59, 62, 67
Alkali, free, 66
Alkali silica reaction (ASR), 12, 59, 67, 86
   diagnosis of, 67
   repairs to damage caused by, 67
Alkaline environments, 11
Alkalis, 61
Ambient temperature, 20
Angle fillets, 44
Anti-friction devices, 40
Appraisal, structural, 62
Arc length, 14
Architects, tasks of, v
Ash,
   black, mortar, 11
Asphalt, 44, 65
   floors, 65
   overheating of, 45
ASR (*see* Alkali silica reaction)
Assembly joints, 19, 21
Atmospheric pollutants, acid, 11
Austenitic stainless steel, 59

Bacteria, 32
Bacteriological decay, 71
Baffle width, 26
Barns, converted, 81
Barriers, vehicle, 81
Beams, 74, 85
   ground, 74
   prestressed, 68
Bed joints, 11, 60, 63, 64
   cracks in, in brickwork, 11
   expansion of, 63
Bituminous felt, 44
Black ash mortar, 11, 59
Blast furnace slag, 67
Boards
   calcium silicate, 66
   fibre, 37
   plaster, 47
Bolted joints, 25
Bomb craters, 71
Bonding, spot, 44

Boundary walls, 37
Bow, 13, 14, 34, 41, 42, 64, 86
   approximate estimation of, 14
   in brickwork, 64
   example of calculation of, 14
   forces related to, 86
   relationship to arc length, 14
Brick gables, 82
Brick rubble, 65
Bricklaying during bad weather, 47
Brickwork, 10, 11, 27, 35, 74, 83
   bed joint cracking in, 11
   bow in, 64
   calcium silicate, 50
   clay, 49
   'green', 39
   moisture expansion of, 27
   moisture-induced size changes in fired clay, 52
   rendered, 63
   repairs to damaged, 64
   sulphate attack in, 10
   unstable, 65
British Geological Survey, 72
Brittleness in roof membranes, 42
BS 8200, example of application of, 39
Builders, tasks of, v
Building(s)
   checklist for site investigations for low rise, 72
   colour of parts of, 33
   complexity, 3
   extensions to existing, 71
   indirect damage to parts of, 82
   inaccuracies in, 21
   integrity, vi
   lack of data for, fabric temperatures, 5
   marketability of, 69
   regulations, vi, 69, 70, 71
   settlement, 75
   tasks of, failure investigators, v, 11, 14, 29
   tasks of, owners, v
   values, v

Cables, electricity, 71
Calcium silicate
   boards, 66
   brickwork, 50
   carbonation of, sheets, 12
   shrinkage in, brickwork, 51
Callipers, digital, 78
Carbon dioxide, atmospheric, 11, 55
Carbonation, 12, 59, 61, 53, 86
   in calcium silicate-based sheets, 12
   in fibre-reinforced cements, 12
Carbonisation of bitumen, 45
Catastrophes, vi
Cathodic protection, 63
Cavity wall ties, 37
Cement, Portland, 63, 86
Cement-based products, 5
   shrinkage of, 47
Ceramic tiles, 53, 58

CRACKING IN BUILDINGS Index

Chalk soils, 70, 71
Changes
    in conditions, 3
    of horizontal size, 35
    in moisture content, 3
    of section, 34, 35
    of size, 3, 35
    of vertical size, 35
Characteristic accuracy, 22, 25
Chases, 34, 39
Checklist for site investigations for low rise building, 72
Chemical reactions, size changes induced by, 10, 59
Chimneys, 10, 63, 82
    sulphate attack in, 10
Chipboard, 56, 57
Chloride content, 63
Chloride ions, 11, 61
Cladding(s) 32, 37, 39, 40, 42
    design principles for, 33
    diagnostic principles for, 41
    high thermal capacity, 39
    low thermal capacity, 39
    practical detailing for, 35
    reflective, 42
    site practices for, 39
    weatherproof, 65
Classification of visible damage to walls, 85
Clay(s)
    brickwork, 49
    floor tiling, 58
    heave in, soils, 65
    map of shrinkable, 76
    reversion of, soils, 77
    seasonal volume changes in, soils, 77, 86
    shrinkable, soils, 74, 77, 87
    slopes, 71
    soils, 70, 74
    zones of swelling in, soils, 74
Removing corrosion products, 61
Clearance, dimensional, 26
Coatings
    metallic, 62
    organic, 62
    surface, 62
Coefficients
    absorption, 39
    of linear thermal expansion, 3, 6, 13
    table of, of linear thermal expansion, 6
    thermal absorption, 37
Cold deck roofs, 43, 44
Collapse, buildings, 2
Colliery shale, 65
Colour of parts of buildings, 33
Columns, reinforced concrete, 27, 37
    shrinkage of, 27
Combustion, spontaneous, 71
Compaction
    concrete, 62
    of fill, 66
Complexity of buildings, 3
Components
    homogeneity of, 3
    lightweight, 20
    protective systems for, 11
    timber, 48
    unrestrained linear response of, 5
Compressible joints, 27
Compressible layers in trench, 73
Compression forces, 86
Compression fractures, 34, 49
Compressive spalling, 19
Concrete
    additives, 11
    blockwork, 50
    compaction, 62
    components, 62
    detachment, 12
    initial shrinkage in, mixes, 10
    lean, mixes, 10
    pore structure of, 67
    reinforced, columns, 27, 37
    spalling, 12
    sulphate attack in, piles, 11
    unrestrained, 67
    variation of drying shrinkage of, 9
    water:cement ratios of, 62
Condensation, 43
Conditions
    changes in, 3
    probable ranges of, 32
Congested reinforcing bars, 62
Consequences of size change, 34
Consolidated ground, 71
Consolidation
    of clay soils, 75
    of granular soils, 75
Contact, edge-to-edge, 25
Contained members, 16
Contained structures, 33, 55
Containing structures, 33, 55
Continuity, structural, 36
Contraction, 40
    moisture-induced, 41
    thermal, 41
Conversion in high alumina cement concrete (HACC), 59, 68
Copings, 10, 12, 37, 38, 65
    sulphate attack in, 10
Core samples, examination of, 12
Cork, 37
Corrosion, 11, 12, 32, 56, 59, 60, 61, 62, 86
    caused by cracking, 62
    of embedded steel in masonry, 60
    inhibitors, 62
    of iron and mild steel, 11
    of metals, 11, 59, 86
    of mild steel reinforcing bars in concrete, 11, 61
    products of, 60
    reinforcement, 11, 67
    of steel, 12
    of strip ties, 64
    volume of, products, 11
    of wall ties, 11, 56, 59, 74
Corrosion-induced cracking, 61
Corrosion-resistant materials, 59
Covermeter, use of, 62
Cracking
    in brickwork bed joints, 11
    caused by explosion, 62

Cracking (cont)
   caused by low temperatures, 44
   in columns, 61
   corrosion-induced, 61, 62
   distinguishing ASR, from initial drying
      shrinkage cracking, 12
   foundation movements caused by, 69
   map-pattern, 12, 54, 55, 65, 67
   mechanism of, 13
   repairs to, 62
   shear, 49, 50, 51, 64
   tensile, 13
   at windows, 51
Crack(s)
   circular pattern, 46
   direction of, 12
   fine, 85
   hairline, vi, 85
   investigation, 86
   linear horizontal, 56
   masking of, 85
   monitoring, 78
   regularly-spaced horizontal, 60
   shear, 51
   shrinkage, 81
   star-shaped, 46
   structural significance of, 32
   tension, 41, 51
   width changes, 77
Craters, bomb, 71
Crazing of tile glaze, 53
Creep, 20, 86
Cross-wall construction, 60
Crystalline deposits, white, 65
Curing
   of renders, 54
   of screeds, 57
Curve, distribution, 22, 25
Cyclical size changes, 5, 41

Daily size changes, 19, 29
Damage
   to existing structures, 71
   frost, 12, 83
   indirect, to parts of buildings, 82
   mechanical, 81
   by punching, 81
   by scraping, 81
   by scuffing, 81
   by trees, 74
   vibration, 81, 82
   to walls, 85
Damaged brickwork, repairs to, 64
Damp proof courses (DPCs), 34, 55, 56, 65, 75, 86
   oversailing of, 10
Damp proof membranes (DPMs), 65, 66
Dampness, persistent, 64
Dead loads, 86
Decay, bacteriological, 71
Deep strip foundations, 70
Defective detailing, 63
Demec points, 80
Desalination, 63
Design, 69
   to minimise size changes, 54

Designers, 33, 59
Desk studies for foundation design, 71
Detachment of concrete, 12
Detritus, 19, 41
Deviation
   Standard, of batch (s), 23
   Standard, of population (σ), 23, 25, 26
De-watering effect, 77
Diagnosis
   of carbonation, 66
   of ground movements, 74
   of HACC conversion, 68
Differential movements, 75, 77
Digital callipers, 78
Direction of cracks, 12
Displacement, 13, 41
Disruption of masonry, 60
Disruptive cracking, 59
Dissimilar materials, proximity of, 50
Distortion
   avoidance of, 14
   of salient corners, 10
   structural, 75
   of walls, 60
Distribution curve, 22, 25
Distribution, normal, 21
Doming of concrete slabs, 11, 65
Door openings, 34
Doors, 51, 85
   cracking at, 51
Dowels, 36
DPCs (*see* Damp proof courses)
DPMs (*see* Damp proof membranes)
Drains, distance from foundations, 70
Droughts, 87
Drying out of new construction, 48
Drying shrinkage
   variation of, of concrete, 9

Eaves
   boxed, 66
   soffits, 66
Edge-to-edge contact, 25
Edge trims, 45
Efflorescent salts, 63
Elastic sealants, 20
Elasticity, modulus of, 13
   table of values of, 15
Elasto-plastic sealants, 20
Electricity cables, 71
Embrittlement of asbestos cement, 66
Environmental temperature change, 33
Environments
   alkaline, 11
   marine, 11
Equilibrium moisture content, wood, 9
Error
   maximum, 24
   random, 21, 26
   systematic, 22, 26
Ettringite, 11, 63, 64, 65
Expansion, 40
   of fill, 65
   joints, 19
   of mortars, 63
   of renders, 63

CRACKING IN BUILDINGS Index

Expansion (*cont*)
   of walls, 74
Expert witnesses, v
Explosion, cracking caused by, 62
Exposure, 5, 33
Extensibility, 42
Extensions to existing buildings, 71

Failure mechanisms, 14
Farmhouses, converted, 81
Felt
   bituminous, 44
   sheathing, 44
Fibre-board, 37
Fibre cement products, 66
Fibre-reinforced cements, carbonation in, 12
Fill (under slabs), 32, 86
   compaction of, 86
   expansion of, 65
Fill, trench, 70, 73
Filler(s)
   compressibility of joint, 36
   joint, 19, 20, 36
   recovery of joint, 19
   strips, 19
   table of movement joint, 20
Fillets, angle, 44
Film, impermeable paint, 12
Fine cracks, 85
Finishes, floor, 57
Fire, 62
Fired clay brickwork, moisture-induced size changes in, 52
Fired clay products, 5, 47
   expansion of, 47
Fixings
   joint, 26
   mechanical, 28
   overdriven, 66
   rigid, 5
Fixity, point of, 41, 43, 45
Flaking (caused by frost), 83
Flat roofs, 32, 42, 43, 44, 45, 82
   design principles for, 42
   diagnostic principles for, 45
   practical detailing for, 43
   site practices for, 44
Flexible sealants, 71, 74
Flooding, 71, 71
Floor(s), 32, 56, 57, 86
   finishes, 57
   gaps between skirtings and, 77
   pitch mastic, 65
   practical detailing for moisture-induced size changes in, 56
   separating, 32
   site practices for installing, 57
   slabs, 86
   in wet areas, 57
Foam plastics, 19
Foam rubber, 37
Footings, sulphate attack in, 11
Forces
   paths of, 13
   rotation, 86
   shear, 86
   tension, 86
Foundation(s)
   desk studies for, design, 71
   lateral loads on, 73
   remedial work to stop, movement, 80
   shallow, 86
   steps in, 69
   strip, 69
Fractures
   compression, 34, 49
   tension, 34, 49
Frames
   structural, 53
   timber, 64
Free alkali, 66
Freestanding walls, 63, 83
   sulphate attack in, 10
Freezing, effects of, 34
Friable mortars, 63
Friction, 13, 28
   restraint, 28
Frost, 12, 34, 62, 70, 71, 83
   damage, 12, 83
   heave, 71
Frost-susceptible soils, 70

Gable end walls, 56
Galvanising, 59
Garages, 71
Gas mains, 71
Geology, 71, 73
Glass, 47, 80
Graticules, 80
'Green' brickwork, 39
Grit blasting, 63
Groove depth, 26
Ground
   beams, 74
   hazardous, 71
   heave, 65, 69, 77, 87
   instability of, 71
   reclaimed, 71
   made, 69, 71
   settlement of, 62, 86, 87
   slopes greater than 1 in 10, 71
   subject to slip or creep, 71
Ground-bearing (-supported) slabs, 11, 74
   sulphate attack in, 11
Groundwater, 32, 71
   sulphate-bearing, 11
Gypsum plasters, 47

HACC (*see* High alumina cement concrete)
Hairline cracks, vi, 85
Hardcore, 66
Hazardous ground, 71
Heave, 65, 69, 77, 87
   in clay soils, 65, 77
   frost, 71
Hedges, 71
Hemp, 37
High alumina cement concrete (HACC), 59, 68
Holes
   fixing, 66
   swallow, 71, 86
Hydration, 59, 61, 66

Hydration (*cont*)
   and carbonation, 66
   of Portland cement, 61

Impacts, 62
Impermeable paint film, 12
Inaccuracies in building, 21
Indirect damage to parts of buildings, 82
Inertia
   thermal, of buildings, 20
Information sources
   on drainage, 72
   on ground conditions, 73
   on topography, 72
   on vegetation, 72
Inhibitors, corrosion, 62
Inspections
   of corroded steel, 62
   suggested approach to, 86
Insulation, 5
Integrity of buildings, vi
Intersecting walls, 34
Intrinsic accuracy, 25
Intrusions, mortar, 39, 40
Inverted roofs, 43
Investigations
   crack, 86
   site, 71
Investigators, tasks of building failure, v, 11, 14, 29
Ions, chloride, 11
Iron, 11, 59
   corrosion of, 11
Irreversible shrinkage, 5
Irreversible size changes, 19, 47

Joint(s)
   and accuracy, 25
   assembly, 21
   bed, 11, 60, 63, 64
   bolted, 25
   compressible, 27
   cracks in bed, in brickwork, 11
   expansion, 19
   expansion of bed, in brickwork, 63
   fillers, 19, 20, 36
   fixings, 26
   location of, in relation to re-entrant corners, 36
   location of, in relation to salient corners, 36
   movement, 19, 34, 53
   perpend, 64
   perimeter, 35
   porous surfaces in, 20
   recovery of, fillers, 19
   riveted, 25
   as safeguards against cracking, 19
   shear, 37, 71
   soft, 19, 52, 39, 53, 54
   spacers, 21
   table of movement, fillers, 20
   welded, 25

Landslips, 73, 86
Lateral loads on foundations, 73
Lateral restraint, 64
Leakage from drains, 75, 86

Lean concrete or mortar mixes, 10
Length of arc, 14
Lift
   motor rooms, 45
   shafts, 43, 45
Lightweight components, 20
Limestone, 71
Linear movements, 78
Linear response
   unrestrained, of components, 5
Linear thermal expansion coefficients, 3, 6, 13
Linings, 85
Lintels, 53
Litigation, 69
Loads
   dead, 86
   lateral, on foundations, 73
   live, 86
   snow, 86
   wind, 86
Location
   of joints in relation to re-entrant corners, 36
   of joints in relation to salient corners, 36
   of restraints, 13
Loss adjusters, tasks of, v
Low-sulphate bricks, 63
Low-temperature cracking, 44

Made ground, 69, 71
MAF (*see* Movement Accommodation Factor)
Magnifiers, use of, 80
Map-pattern cracking, 12, 54, 55, 65, 67
Maps, 71
Marine environments, 11
Marketability of buildings, 69
Masking of cracks, 85
Masonry
   disruption of, 61
   paints, 55
   repairs, 85
Masons, 85
Material(s)
   corrosion-resistant, 59
   painting sheet, 66
   proprietary repair, 63
   proximity of dissimilar, 50
   table of service temperature ranges of, 7
   tensile strength of, 13
   unusual, 86
Maximum error, 24
Mean (of distribution curve), 22, 26
Mean and Standard Deviation, table of, 24
Measurement of levels, 78
Mechanical damage, 81
Mechanical fixings, 13, 28, 40
Mechanisms
   cracking, 13, 55
   examples of cracking, 14
Members, contained, 14
Membranes, 44
   fully bonded, 42
Metallic coatings, 62
Metals, 11, 47
   corrosion of, 11, 59, 86
Methane, 32
Microscopic examination of core samples, 12

CRACKING IN BUILDINGS Index

Mild steel, 59
    corrosion of, 11, 61
Mine shafts, 71
Mining, 71, 73, 86
    records, 73
    subsidence, 86
Misalignment of planes, 41
Mixes, lean concrete or mortar, 10
Modification of size change by restraint, 5
Modulus of elasticity, 13
    table of values of, 15
Moisture content
    changes in, 3
    simultaneous temperature and, changes, 10
    wood equilibrium, 9
Moisture-induced contraction, 41
Moisture-induced size changes, 5, 47, 52, 55, 86
    example of, 9
    table of, 8
Moisture movements, 3
Mortar(s)
    black ash, 11, 59
    expansion of, 63
    friable, 63
    initial shrinkage in, mixes, 10
    intrusions, 39, 40
    lean, mixes, 10
    permeable, 59
Motor rooms, lift, 45
Movement Accommodation Factor (MAF), 20, 34, 49
Movement joint spacing, 49, 50, 51
    in aerated concrete blockwork, 51
    in calcium silicate brickwork, 50
    in fired clay brickwork, 49
    in unreinforced masonry, 35
Movement joint(s), 19, 20, 34, 44, 53
    fillers for, 20
    table of, fillers, 20
Movement(s)
    differential, 75, 77
    linear, 78
    moisture, 3
    remedial work to stop foundation, 80
    reversible, in soils, 77
    rotation, 78
    subsoil, 69
    thermal, 3

Nomogram for thermal expansion, 4
Normal distribution, 21

Occurrence, probability of, 22, 23
    example of, 24
'Oil-canning', 42
Optical probes, 60
Organic coatings, 62
Organic constituents, volatile, 47
Orientation, 5, 33
Overdriven fixings, 66
Overheating of asphalt, 45
Overload, structural, 62
Oversailing, 10, 34, 41, 64
    brickwork, 64
    of damp proof courses, 10
Overtightening fixings, 40

Owners, tasks of building, v

Painting sheet materials, 66
Paint(s)
    impermeable, films, 12
    masonry, 55
    protective, systems, 61
Parapets, 10, 36, 37, 63, 65, 83
    sulphate attack in, 10
Partitions, 86
Patch
    plastering, 74
    repairs, 63
Paths of forces, 13
Peat, 71
Perimeter joints, 35
Permeability to carbon dioxide, 62
Perpend joints, 64
Persistent dampness, 64
PFA (see Pulverised fuel ash)
Phenolphthalein, use of, 67
Photographs for site investigations, 71
Piers, 34
Piles, 70
Pipes, service, 85
Pitch mastic floors, 65
Pits (mines), 71
Pits, trial, 80
Planes
    misalignment of, 41
    slip, 41, 36, 55, 74
Plant machinery bases, 58
Plaster(s)
    boards, 47
    gypsum, 47
    repairs, 85
Plastic sealants, 20
Plastics, 47
    foam, 19
    tell-tales, 80
Plasto-elastic sealants, 20
Point of fixity, 41, 43, 45
Pollutants, acid atmospheric, 11
Polyethylene, cellular, 37
Polyurethane, cellular, 37
Ponds, 71
Population, 23
Pore structure of concrete, 67
Porous jointing surfaces, 20
Portland cement, 63, 86
Pozzolanas, 67
Precompressed strips, 19
Prestressed beams, 68
Prestressing tendons, 67
Probability of occurrence, 22, 23
    example of, 24
Probes, optical, 60
Products
    cement-based, 5
    fired clay, 5
    volume of corrosion, 11
Professional judgement, 80
Proprietary repair materials, 63
Protection, sacrificial, 11
Protective paint systems, 61
Protective systems for components, 11

Protective treatments for steel, 60
PTFE washers, 40
Pulverised fuel ash (PFA), 67
Punching (mechanical damage), 81
Purlins, 82

Quarrying, 71, 73
   records, 73

Radiation
   to clear night sky, 42
   solar, 40, 42
Radon, 32
Rain penetration, 32, 63
Rainwater
   exclusion of, 19
   goods, 63
Random error, 21, 26
Ratchetting effect, 41
Re-alkalisation, 63
Reclaimed ground, 71
Redecoration, 85
Reflective treatments, solar, 44
Refuse tips, 71
Regulations, building, vi
Reinforced concrete columns, 27, 37
Reinforcement
   corrosion of, 11, 67
   depth of cover to, 61
   insecurity of, 62
   non-corrodible, 54
   spacers in, 62
Reinforcing bars, 67
   congested, 62
Relative humidity, 47
Rendered brickwork, 63
Renders, 32, 52, 53, 54, 55, 56, 60, 63
   compressive spalling of, 56
   curing of, 54
   expansion of, 63
   mix proportions for, 54
   specification of, 53
   uninterrupted across joints, 53
   water:cement ratio for, 54
Repairs
   to cracking, 62
   to damaged brickwork, 64
   to damaged slabs, 67
   masonry, 85
   patch, 63
   plaster, 85
   to spalled concrete, 62
Repointing, 74, 85
Re-rendering, 74
Response times, thermal, 33
Restraint(s), 3, 26, 28, 33, 34, 48, 55, 66
   absence of, 55
   absolute, 43
   adhesive, 28
   against bowing, 66
   effectiveness of, 13
   effects of adhesion on, 28
   by fixings, 66
   friction, 28
   lateral, 64
   location of, 13

Retaining walls, 83
Returns, short, 34
Reversible movement in soils, 77
Reversible size changes, 5, 19, 47
Reversion of clay, 77
Ridging, 57
Rigid fixings, 5
Riveted joints, 25
Rock, exposed, 71
Roof(s)
   cold deck, 43, 44
   flat, 42
   membranes, 42
   transfer of, loads, 60
   inverted, 43
   slabs, 86
   warm deck, 43
Roots of trees, 77
Rotation, 34, 41, 78
   in ends of walls, 41
   forces, 86
   movements, 78
Rubber, foam, 37
Rubble, brick, 65
Ruckling, 44, 46
Run (of components), 27
Rust staining, 12, 61, 62

Sacrificial protection, 11
Salts, sulphate, 65, 68
   efflorescent, 63
Samples, examination of core, 12
Sand for render, 54
Sandy soils, 70
Scraping (mechanical damage), 81
Screeds
   curing of, 57
   curling of, 56
   strength of bond to substrate, 57
   water:cement ratio in, 57
Screen walls, 71
Scuffing (mechanical damage), 81
Sealants, 19, 20, 74
   elastic, 20
   elasto-plastic, 20
   flexible, 71, 74
   plastic, 20
   plasto-elastic, 20
   width-to-depth ratios of, 20, 36
Seasonal size changes, 19, 29
Seasonal temperature ranges, 37
Seasonal volume changes in clay soils, 77, 86
Section, changes of, 34, 35
Separating floors, 32
Separating walls, 32
Service pipes, 85
Service temperature ranges of materials, table of, 7
Service trenches, 70
Settlement
   of buildings, 75
   of ground, 62, 86, 87
Sewers, 71
Shading, solar, 5, 42
Shafts, lift, 43, 45
Shale, colliery, 65

Shallow foundations, 86
Shear, 34, 49, 50, 51, 64, 71, 86
    cracking, 49, 50, 51, 64
    forces, 86
    joints, 37, 71
Sheathing felt, 44
Sheet materials, painting, 66
Short returns, 34
Shrinkable clays, 74, 77, 87
    map of, 76
Shrinkage
    of cement-based products, 47
    cracks, 81
    initial, in concrete and mortar mixes, 10
    irreversible, 5
    of reinforced concrete columns, 27
    of substrate before tiling, 55
    of timber, 53
Siliceous aggregates, 67
Sills
    sulphate attack in, 10
    window, 53, 54
Site investigations
    checklist for, for low rise building, 72
    photographs for, 71
Site practices, 47
Site supervision, 40
Size change(s), 3, 35
    chemically-induced, 10, 59
    cyclical, 5, 41
    daily, 19, 29
    design to minimise, 54
    example of moisture-induced, 9
    example of temperature-induced, 5
    examples of, in calcium silicate brick walls, 36
    examples of, in clay brick walls, 36
    examples of, in concrete block walls, 36
    examples of unrestrained, in brick and block walls, 36
    expansive, 42
    interdependency of causes of, 10
    irreversible, 19, 47
    modification of, by restraints, 5
    moisture-induced, 5, 47, 52, 55, 86
    mutually compensating, 48
    rapid, 42
    reversible, 5, 19, 47
    seasonal, 19, 29
    simultaneous temperature and moisture content, 10
    table of moisture-induced, 8
    temperature-induced, 3, 33
    thermally-induced, 3
Size
    target, of components, 26, 27
Skirtings, gaps between floors and, 77
Slabs
    doming of concrete, 11, 65
    floor, 86
    ground-bearing (-supported), 11, 32, 65, 74
    roof, 86
Slag, blast furnace, 67
Slip planes, 36, 41, 55, 74
Slips, disruption of brick, 52
Slope angles, 71
Slots for adjustment, 28, 29, 40

Snow loads, 86
Soft joints, 19, 39, 52, 53, 54
    omission of, 54
    tiling over, 54
Soil(s)
    clay, 65, 70, 74, 77
    frost-susceptible, 70
    heave in clay, 65
    map of shrinkable clay, 76
    reversible movement in, 77
    sandy, 70
    seasonal volume changes in clay, 77
    shrinkable clay, 74, 77, 87
    types of, 70, 86
    zone of swelling, 74
Solar radiation, 40, 42
Solar reflective treatments, 44
Solar shading, 5, 42
Soldier courses, sulphate attack in, 10
Sound transmission, airborne, 32
Sources of information
    on drainage, 72
    on ground conditions, 73
    on topography, 72
    on vegetation, 72
Spacers, 19, 21, 39
    joint, 21
    reinforcement, 62
    timber, 39
Spalling, 12, 19, 42, 61, 62, 83
    compressive, 19
    concrete, 12
    of renders, 56
    repairs, 62
Specifiers, tasks of, 59
Spontaneous combustion, 71
Spot bonding, 44
Stables, converted, 81
Staining, rust, 12, 61, 62
Stainless steel, austenitic, 59
Standard Deviation
    of batch (s), 23
    of population (σ), 23, 25, 26
    table of, 24
Steel
    austenitic stainless, 59
    corrosion of, 11, 12
    corrosion of embedded, in masonry, 60
    protective treatments for, 60
    use of, rules, 80
Steel-framed structures, 32
Steps in foundations, 69
Stirrups, 12
Straight-edges, use of, 46, 65
Strain, 13
    tensile, 20
    localised, 42, 44
Strength
    tensile, of material, 13
Stress, 13
    evaluation of, 13
    tolerable, 13
Strip foundations, 69
Strip ties, corrosion of, 64
Strips
    filler, 19

Strips (*cont*)
   precompressed, 19
Structural appraisal, 62
Structural connections, 60
Structural continuity, 36
Structural distortion, 75
Structural frames, 53
Structural members, creep in, 20
Structural overload, 62
Structural serviceability, 59
Structural significance of cracks, 32, 69
Structural tying, 37
Structural weakness, 75
Structures
   contained, 33, 55
   containing, 33, 55
   damage to existing, 71
   steel-framed, 32
Subsidence, 62, 69, 87
Subsoil movements, 69
Substrate
   shrinkage of, before tiling, 55
   change of, 53
Sulphate attack, 32, 54, 55, 56, 59, 60, 63, 64, 66, 74, 75, 82, 86
   in brickwork, 10
   in chimneys, 10
   in concrete piles, 11
   in copings, 10
   cracks in bed joints caused by, 65
   diagnosis of, 63
   distortion of salient corners, 10
   in footings, 11
   in freestanding walls, 10
   in ground-bearing (-supported) slabs, 11, 65
   oversailing of DPCs, 10
   in parapets, 10
   in sills, 10
   in soldier courses, 10
Sulphate-bearing groundwater, 11
Sulphate-resisting Portland cement, 63
Sulphate salts, 65, 68
Sulphides, 59
Supervision, site, 40
Surface coatings, 62
Surfaces, porous jointing, 20
Surveys, walk-over, 71
Surveyors, tasks of, v, 11, 29, 14, 33, 59, 60, 68, 81
Swallow holes, 71, 86
Sway, 86
Systematic error, 22, 26
Systems
   protective, for components, 11

Table
   of modulus of elasticity, 15
   of movement joint fillers, 20
   of thermal expansion coefficients, 6
   of tree root problems, 79
Tanking, 32, 57
Target size of components, 26, 27
Tell-tales
   glass, 80
   plastics, 80

Temperature(s)
   ambient, 20
   change of environmental, 33
   cracking caused by low, 44
   at installation, 39
   lack of data for, of building fabric 5
   prevailing, 34
   seasonal range of, 37
   simultaneous, and moisture content changes, 10
Temperature-induced size changes, 3, 33, 86
Tensile cracking, 13
Tensile strain, 20
Tensile strength of material, 13
Tension cracks, 41, 51
Tension forces, 86
Tension fractures, 34, 49
Tension widths, 74
'Tenting', 57
Thaumasite, 11, 63, 65
Thermal absorption coefficients, 37
Thermal capacity of cladding, 39
Thermal contraction, 41
Thermal expansion
   coefficients of linear, 3, 6, 13
   computation of, 3
   nomogram for, 4
   table of, coefficients, 6
Thermal inertia of construction, 20
Thermal movements, 3
Thermal response times, 33
Thermally-induced size changes, 3
Thin sheet products, 66
Ties
   cavity wall, 37
   corrosion of wall, 11, 56, 74
Tiled walls, 54
Tiles, ceramic, 53, 58
Tiling
   clay floor, 58
   over soft joints, 54
Timber
   components, 48
   frames, 64
   shrinkage of, 53
   spacers, 39
Timber-framed walls, 53
Tips, refuse, 71
Tolerable stress, 13
Tolerances, 21, 22, 24
Toothing-in, 71
Total Relevant Movement (TRM), 20
Toxic wastes, 71
Trees, 71, 77, 80, 86, 87
   on clay soils, 75
   damage by, 74
   felled, 77
   removal of, 73
Tree root problems, table of, 79
Trench fill, 70, 73
Trenches
   compressible layers to, 73
   service, 70
Trial pits, 80
Trims, edge, 45

CRACKING IN BUILDINGS Index

TRM (*see* Total Relevant Movement)
Tying, structural, 37
Types of soil, 70, 86

Underpinning, 78, 80
Unrestrained concrete, 67
Unrestrained linear response of components, 5
Unstable brickwork, 65
Unusual materials, 86

Values, building, v
Values of modulus of elasticity, 13
Vapour control layers, 32, 43
Variation of drying shrinkage of concrete, 9
Vegetation, 71, 77
   influence of, 77
Vehicle barriers, 81
Vehicles, 81
Verges, 53
Vibration damage, 81, 82
Volatile organic constituents, 47
Volume changes in clay soils, seasonal, 77, 86
Volume of corrosion products, 11

Walk-over surveys, 71
Wall ties
   corrosion of, 11, 56, 74
   diagnosis of corrosion of, 59
   protection of, 59
   related to age of building, 59
   sampling of, 60
   vertical twist, 59
   wire, 60
Walls
   abrasion of, 82
   boundary, 37
   damage to, 85
   design principles for, 33, 48
   diagnostic principles for, 41, 55
   distortion of, 60
   expansion of, 74
   freestanding, 63, 83
   gable end, 56
   intersecting, 34
   practical detailing for, 35, 51
   retaining, 83
   rotation of ends of, 41, 46
   screen, 71
   separating, 32
   site practices for, 39, 53
   sulphate attack in freestanding, 10
   tiled, 54
   timber-framed, 53
Warm deck roofs, 43
Washers, PTFE, 40
Wastes, toxic, 71
Water:cement ratios, 62, 68
   in screeds, 57
   of concrete, 62
Water mains, 71
Water tables, 70, 71
Watercourses, buried, 71
Weakness, structural, 75
Weatherproof claddings, 65
Weathertightness, 85
Welded joints, 25

Wells, 71
White crystalline deposits, 65
Width-to-depth ratios of sealants, 20, 36
Wind loads, 86
Window(s), 34, 51, 53, 85
   cracking at, 51
   openings, 34
   sills, 53, 54
Witnesses, tasks of expert, v
Wood block floors
   displacement of, 57
   finishes for, 57
Wood equilibrium moisture content, 9
Wood panel finishes, 57
Worth of buildings, v

Zones of swelling, 74